Biotin and Other Interferences in Immunoassays

A Concise Guide

Biotin and Other Interferences in Immunoassays

A Concise Guide

AMITAVA DASGUPTA, PHD

Department of Pathology and Laboratory Medicine
University of Texas McGovern Medical School at Houston
Houston, TX

ELSEVIER

ELSEVIER

3251 Riverport Lane
St. Louis, Missouri 63043

Biotin and Other Interferences in Immunoassays   ISBN: 978-0-12-816429-7

Publisher: Stacy Masucci
Acquisition Editor: Tari K. Broderick
Editorial Project Manager: Megan Ashdown
Production Project Manager: Sreejith Viswanathan
Cover Designer: Miles Hitchen

Printed in United States of America

Preface

Biotin, known as vitamin B_7 or vitamin H ("H" is the initial for "Haar" and "Haut" the German word for hair and skin, respectively), is a member of vitamin B complex and is water soluble. Biotin was first isolated in 1936 as a yeast growth factor from egg yolk by Kogl and Tonnis and the structure was elucidated in 1942 by du Vigneaud et al. Biotin is essential for normal cellular functions, growth, and development. Humans and other mammals cannot synthesize biotin and must obtain biotin from diet via intestinal absorption. However, normal microflora of the large intestine can also synthesize biotin. The daily requirement of biotin is only 30 µg but in recent days, people take high doses of biotin (1–10 mg/day) for healthy hair, nails, and skin. People with rare genetic biotin deficiency may need 40–100 mg biotin per day, while preliminary results from clinical trials show that 100–300 mg biotin per day may be effective in reducing various symptoms of multiple sclerosis. Biotin is water soluble and relatively nontoxic.

Immunoassays are widely used in clinical laboratories for rapid turnaround of results. However, immunoassays are affected by both endogenous and exogenous substances producing inaccurate (both falsely elevated and falsely lower) results. Although interferences in immunoassays are due to structurally related compounds, such as interference of a drug metabolite in the analysis of the parent drug, sometimes structurally unrelated compounds may cause significant interference in immunoassays. One of such interferences is biotin interference in immunoassays that utilize biotinylated antibodies and streptavidin-coated magnetic beads as a means of immobilizing antigen-antibody complexes to the solid phase. This design offers many advantages including signal amplification and increased sensitivity. In the recent years, taking high doses of biotin supplement is gaining popularity among the general population and as a result, interference of biotin in biotin-based immunoassays (false-positive results in competitive immunoassays but false-negative results in sandwich assays) is becoming a serious clinical issue causing wrong diagnosis. The most commonly reported problem is misdiagnosis of Graves disease due to falsely lower thyroid-stimulating hormone levels (sandwich immunoassay format) but falsely elevated free triiodothyronine and free thyroxine levels (both assays use competitive format) due to biotin interference in thyroid function tests using immunoassays that utilize biotinylated antibodies. FDA reported the death of a person due to missed diagnosis of myocardial infarction due to falsely lower troponin value in the person taking biotin supplement.

At present, biotin interferences have been reported in clinical chemistry and therapeutic drug monitoring tests using immunoassays. This topic is discussed in detail in this book including current approaches of minimizing or eliminating such effects (Chapter 4). In addition, pharmacology and physiology (Chapter 2), as well as use of biotin supplement, are also addressed in the book (Chapter 3). However, in order to have a complete guide for laboratory scientists, pathologists, and clinicians, other interferences in immunoassays used in testing various analytes in clinical chemistry and toxicology laboratory are also addressed in this book (Chapters 5–7) so that readers can use the book for familiarizing with all sources of interferences, including biotin interferences in clinical laboratory tests. I would like to thank my wife for her support during the long evening and weekend hours I spent in writing this book. If readers enjoy reading this book, my effort will be dully rewarded.

Respectfully submitted by
Amitava Dasgupta
Houston, TX

Contents

CHAPTER 1

Immunoassay Design and Mechanism of Biotin Interference

INTRODUCTION

Immunoassays are used in clinical laboratories for analysis of a variety of analytes including hormones, serum proteins, antibodies to infectious or allergic agents, therapeutic drugs that need monitoring, and drugs of abuse testing. These immunoassays exhibit high sensitivity and broad analytical range. Most immunoassay methods use specimens without any pretreatment, and such assay can be easily run on fully automated, continuous, high-throughput, random-access systems. However, for a few immunoassays (cyclosporine, tacrolimus, sirolimus, and everolimus) extraction of the analyte from whole blood is needed prior to the analytical step; however, the cyclosporine, tacrolimus, and sirolimus assays on the Dimension analyzer platform marketed by Siemens require no manual specimen pretreatment because pretreatment is fully automated in the analyzer.

A major advantage of immunoassays is the small specimen requirements (10–300 μL of serum or plasma). Moreover, reagents may be stored in the analyzer, most have stored calibration curves on the automated analyzer system; reagents are often stable for a few weeks to 1 month; and the results can be reported in 10–30 min. Immunoassays offer fast-throughput analysis, automated rerun, autoflagging (to alert for poor specimen quality due to conditions such as hemolysis, lipemia, or high bilirubin content), and high sensitivity and specificity, and results can be reported directly to the laboratory information system.

IMMUNOASSAY DESIGN

Immunoassays measure the analyte concentration in a specimen by forming a complex with a specific binding molecule, which in most cases is an analyte-specific antibody (or a pair of specific antibodies). However, immunoassays are a form of macromolecular binding reaction because no covalent chemical bond is formed between the assay antibody and the antigen molecule (assay analyte). The interaction between antibody and antigen is due to weak hydrogen bonding and van der Waals forces. In general, the Fab region of the antibody is responsible for antibody-antigen interactions. Almost all antibodies used in immunoassays are immunoglobulin (Ig) G antibodies. The N-terminal 110 amino acid residues of the both heavy and light chains of Ig molecules are variable in sequence and interact with antigen to form antigen-binding sites. It has been speculated that 30 amino acids are involved in antigen contact ratio but X-ray crystallographic studies have shown that a maximum of only 17 amino acids are involved in binding of antigen to antibody. However, variation in these 17 amino acid sequences results in a vast number of antibodies capable of binding a wide range of antigens (analytes) with high specificities.[1] Therefore, immunoassay is a bioanalytical technique that is based on the avidity and specificity of antigen-antibody reaction.

Immunoassays can be broadly classified under two categories:
- Competitive immunoassays
- Noncompetitive, also known as immunometric or sandwich immunoassays

Competitive immunoassays utilize only a single antibody against a specific analyte (antigen) and are useful for analysis of small molecules, including various therapeutic drugs and drugs of abuse. However, sandwich immunoassays require two antibodies recognizing two different sites of the analyte molecule and are utilized for assay of large molecules, such as thyroid-stimulating hormone (TSH), cardiac troponin T, cardiac troponin I, various tumor markers, peptides.

Another way of classifying immunoassay format is
- homogeneous immunoassay
- heterogeneous immunoassay

In homogeneous immunoassay, after an antigen-antibody complex is formed, separation of bound antigen (or bound labeled antigen) and free antigen is not required prior to measuring assay signal. However, in the heterogeneous format, bound form must be separated from the free form usually by incorporating a washing step prior to measuring the assay signal. In the heterogeneous assay format the antibody is often

Biotin and Other Interferences in Immunoassays. https://doi.org/10.1016/B978-0-12-816429-7.00001-0

immobilized on a solid support such as microparticle, whereas homogeneous immunoassay takes place in a solution.[2]

In competitive immunoassays the analyte in the specimen competes with the labeled analyte for limited antibody-binding sites and the signal is measured either without separation (homogeneous format) or after separation (heterogeneous format) of bound labeled antigen-antibody complexes from free labeled antigen-antibody complexes . In competitive immunoassay, after the reaction is complete, either labeled antigen bound to the antibody generates the signal or labeled unbound antigen generates the signal.

In sandwich immunoassays the analyte in the specimen binds to two different antibodies that recognize two separate binding sites in the antigen molecule (different epitopes); one antibody may be conjugated to a solid phase and the other to a label. The bound complex is separated from other assay components by a proper washing protocol, and the relevant amount of label produces the signal. The signal in sandwich assays is directly proportional to the analyte concentration with low background noise. Therefore, this type of immunoassay can be highly sensitive, capable of detecting very low concentrations of the analyte. The signals generated are mostly optical—absorbance, fluorescence, or chemiluminescence.

IMMUNOASSAY REAGENTS

The main reagent in any immunoassay is the binding molecule, which is most commonly an analyte-specific antibody or fragments of the antibody, generated by digestion of the antibody by peptidases, e.g., Fab, Fab′ (or their dimeric complexes). Both polyclonal and monoclonal antibodies are used in immunoassays. Polyclonal antibody is raised in animals such as rabbit, goat, or sheep (or other animals) where the analyte (as antigen) along with an adjuvant is injected into the animal. Low-molecular-weight analytes are most commonly injected as conjugates to a large protein, for example, bovine serum albumin. Appearance of analyte-specific antibodies from different B cells of the animal that recognize different epitopes of the same antigen in the animal's sera is monitored, and when a sufficient concentration of the antibody is reached, the animal is bled. The serum can directly be used as the analyte-specific binder in an immunoassay; however, in most cases, antibodies are purified from serum prior to use in immunoassays. As there are many clones of the antibodies for the analyte, these antibodies are called polyclonal.

However, monoclonal antibodies are gaining popularity in immunoassays design owing to their improved analyte specificity. In contrast to polyclonal antibodies, which are generated by different B cells, monoclonal antibodies are produced by a single line of B cells. Monoclonal antibody is produced using the hybridoma technology (hybrids between myeloma cells and antibody-producing cells) first described by Kohler and Milstein. In this technology, immortal myeloma cells are fused with antibody-producing B lymphocytes using polyethylene glycol to break down cell membranes and allowing mixing of genetic materials from both cell types. The new hybridoma cell line is a fusion between myeloma cells and B cells.[3] However, the first step of producing monoclonal antibody is the same as producing polyclonal antibody where an animal, most commonly mouse, is injected with the antigen of interest but the spleen cells rather than the serum are removed from the animal at a carefully defined time subsequent to a final booster immunization. Then single B-lymphocytes generating antibodies to one specific epitope of the antigen can be isolated from the spleen cells and lymph nodes of the immunized animals. However, these cells have limited life span producing only limited amount of antibody. In order to increase antibody production, these cells are fused with immortal myeloma cells to produce the hybridoma cell line, which grows in a medium containing aminopterin (to kill unfused parent myeloma cells). Then hybridoma grows uncontrollably producing only the single clone of desired antibody, known as monoclonal antibody present in the supernatant.[4]

There are several benefits of monoclonal antibodies over polyclonal ones:

- The characteristics of polyclonal antibodies are dependent on the animal producing the antibodies; if the source animal must be changed, the resultant antibody may be quite different.
- Polyclonal antibodies constitute many antibody clones and may be less specific than monoclonal antibodies. However, polyclonal antibodies may be useful for capturing large molecules, such as proteins, in the sandwich assay format.

The other main reagent component of the immunoassay is the label. There are many different kinds of labels generating different kinds of signals. For example, use of acridinium ester labels, when treated with peroxide, produces chemiluminescent signals. As described earlier, an enzyme may be used as the label, which can generate different types of signals depending on the substrate used for the enzyme.

COMMERCIALLY AVAILABLE HOMOGENEOUS IMMUNOASSAYS

The first generation of commercially available immunoassays (i.e., radioimmunoassays) was developed after the landmark discovery of immunoassays by Solomon Berson and Rosalyn Yalow.[5] Immunoassays are widely used in clinical laboratories for measuring more than 100 analytes. The global immunoassay market was estimated at 17.16 billion in 2017.[6] Many diagnostic companies market immunoassays for application in clinical laboratories using various formats, which are discussed in this section. The various types of commercially available immunoassays are summarized in Table 1.1.

Although radioimmunoassays were extensively used in clinical laboratories in the past, the major problem with radioimmunoassays was the use of radioactive material that requires extra precautions to protect laboratory personnel from exposure to radioactive materials. Moreover, special steps must also be adopted for disposal of radioactive materials. The health concerns and high costs of waste disposal make radioimmunoassays burdensome to use. As a result, enzyme-based immunoassays were developed to overcome limitations of radioimmunoassays. In enzyme immunoassays, an enzyme replaces radioactive iodine as the tracer. Today,

both enzyme-based and non-enzyme-based immunoassays are available commercially. In nonenzymatic immunoassays, a label that may generate fluorescent or chemiluminescent signal replaces radioactive iodine in the assay design. In this section, commercially available homogeneous assays are discussed.

Fluorescence Polarization Immunoassays

Fluorescence polarization immunoassay (FPIA), originally developed and marketed by the Abbott Laboratories, utilizes a small fluorescent molecule as the label, which is attached to the analyte molecule. This label has different Brownian motion when it is in the free form versus complexed to a large antibody. The FPIA is a homogeneous competitive immunoassay where after the incubation, fluorescence polarization signal is measured without separation of bound labels from free labels. If the labeled antigen is bound to the antibody molecule, then signal is generated because upon binding of the fluorescent molecule to a much larger molecule, such as an antibody, the degree of polarization increases due to the reduced rotation of the bound fluorescent tracer. In contrast, when the labeled antigen is free in the solution, it can rotate freely (more Brownian motion) thus producing no signal. Therefore, intensity of the signal is inversely proportional to the analyte

TABLE 1.1
Various Types of Commercially Available Immunoassays

Assay Format	Example	Signal	Application	Comments
Competitive/ Homogeneous	EMIT	Absorption at 340 nm due to conversion of NAD to NADH	Therapeutic drug, drugs of abuse, and other analytes	First introduced by Syva but now used in multiple platforms such as ADVIA Chemistry, Viva, and Dimension analyzer (Siemens), as well as in platforms marketed by Beckman Coulter. Some DRI assays marketed by Thermo Fisher are also based on EMIT.
Competitive/ Homogeneous	FPIA	Fluorescence polarization	Therapeutic drugs and drugs of abuse	First introduced by Abbott Laboratories but this format has been discontinued by Abbott. Moreover, assays can only be run using TDx and AxSYM analyzers. The TDx analyzer is no longer available in the United States.
Competitive/ Homogeneous	CEDIA	Colorimetric at visible wavelength (570 nm)	Therapeutic drugs, drugs of abuse, and other analytes	Colorimetry (enzyme modulation, commonly using β-galactoside enzyme fragments) at a system-dependent visible wavelength. Many assays marketed by Thermo Fisher, such as assays for therapeutic drugs, drugs of abuse, thyroid function tests, folate, and vitamin B_{12}, are based on CEDIA.
Competitive/ Homogeneous	KIMS	Colorimetric at various visible wavelengths (500–650 nm)	Drugs of abuse	The On-Line Drugs of Abuse Testing immunoassays marketed by Roche Diagnostics (Indianapolis, IN) are based on the KIMS format.

Continued

TABLE 1.1
Various Types of Commercially Available Immunoassays—cont'd

Assay Format	Example	Signal	Application	Comments
Competitive/ Homogeneous	TIA	Turbidimetry or nephelometry	Various analytes	In TIA, analytes (antigen) or their analogs are coupled to colloidal particles (made of latex). As antibodies are bivalent, the latex particles agglutinate in the presence of the antibody. However, in the presence of free analyte in the specimen, there is less agglutination and the resulting turbidity can be monitored as the end point or as the rate. Siemens and Roche Diagnostics use this technique for analysis of various serum proteins.
Homogeneous	LOCI	Colorimetry, visible wavelength at 612 nm	Various analytes	LOCI can be used for both small and big molecules. For detection of small molecules the assay format is competitive, but for large molecules the sandwich assay format is used. LOCIs utilize biotinylated antibodies. Therefore, high biotin concentration in serum/plasma may falsely elevate analyte concentration for small analytes (competitive format) and falsely lower the true analyte concentration for large molecules (sandwich format).
Heterogeneous	CMIA	Chemiluminescent reaction is measured as relative light units	Various analytes	CMIA is a two-step immunoassay that can be used for the analysis of both small molecules and large molecules.
Heterogeneous	ECL immunoassay	Application of voltage to electrodes induces chemiluminescent reaction that is measured by a photomultiplier	Various analytes	ECL immunoassay can be used for analysis of small molecules (competitive format) as well as large molecules (sandwich format). This design is used in the Elecsys automated immunoassay system from Roche Diagnostics. However, ECL assays from Roche utilize biotinylated antibodies. Therefore, a high biotin concentration in serum may falsely elevate the true analyte concentration for small molecules (competitive format) and falsely lower the true analyte concentration for large molecules.
Heterogeneous/ Competitive	ACMIA	Colorimetry, visible wavelength at 577 and 700 nm	Therapeutic drugs	Cyclosporine, tacrolimus and sirolimus assay marketed by Siemens is based on ACMIA technology, and it is the only commercially available immunoassay requiring no specimen pretreatment.
Heterogeneous/ Sandwich	ELISA	Chemiluminescent, fluorometric, or colorimetric detection, depending on the assay design	Various analytes	Examples of commercial ELISA-using chemiluminescent labels are Immulite (Siemens) and ACCESS from Beckman Coulter.

ACMIA, antibody-conjugated magnetic immunoassay; *CEDIA*, cloned enzyme donor immunoassay; *CMIA*, chemiluminescent microparticle immunoassay; *ECL*, electrochemiluminescent; *ELISA*, enzyme-linked immunosorbent assay; *EMIT*, enzyme-multiplied immunoassay technique; *FPIA*, fluorescence polarization immunoassay; *KIMS*, kinetic interaction of microparticles in solution; *LOCI*, luminescent oxygen channeling assay; *NAD*, nicotinamide adenine dinucleotide; *TIA*, turbidimetric immunoassay.

concentration.[7] In the past, Abbott Laboratories marketed many assays for therapeutic drug monitoring and drugs of abuse testing utilizing FPIA design.[8] However, FPIAs require special analyzers such as TDx and AxSYM, which further restricts the application of FPIA, particularly in automated clinical laboratories. In 2011, Abbott Laboratories withdrew TDx analyzer from the US market.

Enzyme-Multiplied Immunoassay Technique
Enzyme-multiplied immunoassay technique (EMIT) is one of the first homogeneous enzyme immunoassays available commercially, originally developed by the Syva Company in 1973. In EMIT, the label enzyme, glucose 6-phosphate dehydrogenase, is active unless the labeled antigen is bound to the antibody when it loses its enzymatic activity due to steric hindrance. The active enzyme reduces nicotinamide adenine dinucleotide (NAD) to NADH, and the absorbance is monitored at 340 nm (NAD has no absorption at 340 nm but NADH absorbs at 340 nm). In EMIT, intensity of the signal is directly proportional to the analyte concentration because if high concentration of analyte is present in the specimen then few labeled antigen would be able to bind with limited antibodies present in the reaction mixture. Therefore, at equilibrium, most labeled antigen should be free in the reaction mixture retaining enzymatic activity and more NAD would be converted to NADH.[7] EMIT assays are commercially available for analysis of both many therapeutic drugs using serum/plasma specimens and drugs of abuse using urine specimens. Currently, many drugs of abuse assays marketed by Siemens utilize EMIT.[9]

Cloned Enzyme Donor Immunoassay
The cloned enzyme donor immunoassay (CEDIA) method is based on recombinant DNA technology to produce a unique competitive homogeneous enzyme immunoassay format. The assay principle is based on the bacterial enzyme β-galactosidase, which has been genetically engineered into two inactive fragments. The small fragment is termed as enzyme donor (ED) that can freely associate in solution with the larger part of the enzyme known as enzyme acceptor (EA). When both fragments of the enzyme are combined, an intact enzyme is produced with enzymatic activity capable of cleaving a substrate, thus generating a color change in the medium. This color change can be measured spectrophotometrically. In this assay, analyte molecules in the specimen compete for limited antibody-binding sites with analyte molecules conjugated with the ED

fragment. The covalent attachment of analyte to ED does not affect the ability of EA and ED to form active enzyme. If analyte molecules are present in the specimen, then they bind to antibody-binding sites leaving analyte molecules conjugated with ED free to form active enzyme by binding with EA, and then the active intact enzyme generated can be monitored through hydrolysis of an appropriate substrate such as chlorophenol red-β-D-galactopyranoside. The intensity of the signal is directly proportional to the analyte concentration and is usually measured in the visible wavelength at 570 nm. Many therapeutic drugs of abuse and assays for other analytes manufactured by Microgenics Corporation (a subsidiary of Thermo Fisher) utilizes the CEDIA format, and other commercial assays are also based on this format.[10]

Kinetic Interaction of Microparticles in Solution
In the kinetic interaction of microparticles in solution (KIMS) format, in the absence of antigen (analyte) molecules, free antibodies bind to drug microparticle conjugates forming particle aggregates that result in an increase in absorption, which is optically measured at various visible wavelengths (500–650 nm). When antigen molecules are present in the specimen, antigen molecules bind with free antibody molecules and prevent formation of particle aggregates, resulting in diminished absorbance in proportion to the drug concentration. The On-Line Drugs of Abuse Testing immunoassays marketed by Roche Diagnostics (Indianapolis, IN) are based on the KIMS format.

Turbidimetric Immunoassays
Turbidimetric immunoassays use homogeneous immunoassay format where analytes (antigen) or their analogs are coupled to colloidal particles. As antibodies are bivalent, the latex particles agglutinate in the presence of the antibody. However, in the presence of free analyte in the specimen, there is less agglutination and the resulting turbidity can be monitored as the end point or as the rate.

Luminescent Oxygen Channeling Assay (LOCI)
Luminescent oxygen channeling assay (LOCI) is a homogeneous competitive immunoassay capable of rapid quantitative determination of a wide range of analytes in low to high concentrations. A unique feature of this homogeneous immunoassay is that it is capable of analyzing both small molecules, such as

therapeutic drugs, and relatively large molecules, for example, TSH, myoglobin, creatine kinase-MB (CK-MB). For small molecules, one antibody (competitive format) is used, but for large molecules the sandwich format is used. This technology is used in the Siemens Dimension Vista automated assay system. For example, in the LOCI digoxin assay (Siemens), the first bead reagent (Chemibeads) is coated with ouabain, a weaker binding analog of digoxin and also contains a photosensitizer dye. In the first step, the specimen is incubated with a biotinylated F(ab')2 fragment of an antidigoxin mouse monoclonal antibody which allows digoxin from serum/plasma to saturate a fraction of the biotinylated antibody that is directly related to the digoxin concentration. In the second step, ouabain-coated Chemibeads are added to form a biotinylated F(ab')2 immune complex with the nonsaturated fraction of the biotinylated F(ab')2 antibody. Finally, streptavidin-coated Sensibeads are added to bind to biotin to form bead-pair complex. Illumination of the complex at 680 nm generates singlet oxygen species from Sensibeads, which diffuse to Chemibeads and produce a chemiluminescent reaction. The resulting signal is measured at 612 nm and the signal is inversely proportional to serum digoxin concentration.

For analysis of cardiac troponin I, CK-MB, myoglobin, and other large molecules, the LOCI technology utilizes two antibodies and the sandwich assay format. For example, in the LOCI cardiac troponin I assay, two synthetic bead reagents and a biotinylated anti–cardiac troponin I antibody fragment are used. Sensibeads are coated with streptavidin and contains a photosensitizer dye, while Chemibeads are coated with a second anti–cardiac troponin I monoclonal antibody and contain a chemiluminescent dye. The specimen containing cardiac troponin I is incubated with Chemibeads and biotinylated antibodies to form bead-cardiac troponin I–biotinylated antibody sandwich. Then Sensibeads are added to bind biotin to form bead pair immune complexes. When such complexes are illuminated at 680 nm, singlet oxygen species from Sensibeads diffuse to Chemibeads yielding a chemiluminescent reaction. The signal measured at 612 nm is directly proportional to serum cardiac troponin I concentration. Kelley et al.[11] reported that the LOCI assay for cardiac troponin I, CK-MB, myoglobin, and N-terminal prohormone brain natriuretic peptide (NT-proBNP) for the application of the Dimension Vista analyzer demonstrated acceptable performance characteristics for use as an aid in the diagnosis and risk assessment of patients presenting with suspected acute coronary syndromes.

The LOCI TSH homogeneous chemiluminescent immunoassay (CLIA) manufactured by Siemens also involves the use of two latex bead reagents and a biotinylated monoclonal antibody fragment. But this assay is a sandwich immunoassay because TSH is a large molecule. However, the LOCI assay for measuring free thyroxine (FT_4) and free triiodothyronine (FT_3) utilizes only one monoclonal biotinylated antibody (anti-T_4 and anti-T_3, respectively) and competitive assay format because both FT_4 and FT_3 are small molecules. Chemibeads are coated with T_3 and diiodothyronine (T_2) and according to the principle of competitive binding; the final chemiluminescence signal is inversely related to the free thyroid hormone concentration.[12] Other studies have also shown usefulness of LOCI assays for measuring various analytes in clinical laboratories.[13] Patel et al.[14] applied the LOCI technology for quantification of DNA.

COMMERCIALLY AVAILABLE HETEROGENEOUS IMMUNOASSAYS

In heterogeneous immunoassays the bound label is physically separated from the unbound label prior to measuring the signal. The separation is often done magnetically using paramagnetic particles, and after separation of bound from free labels by washing, the bound label is reacted with other reagents to generate the signal. This is the mechanism in many CLIAs, where the label may be a small molecule that generates a chemiluminescent signal. Examples of immunoassay systems where the chemiluminescent labels generate signals by chemical reaction are the ADVIA Centaur from Siemens,[15] Abbott Architect, and some assays manufactured by Beckman Coulter. CLIAs are commonly used for analysis of hormones and serum proteins.

Chemiluminescent Microparticle Immunoassay

The chemiluminescent microparticle immunoassay (CMIA) assay technique can be applied for analysis of both small molecules and large molecules. For example, in the cyclosporine CMIA marketed by Abbott Laboratories for application on the Architect i analyzer is a two-step immunoassay. After extraction of cyclosporine from whole blood using the reagent supplied by the manufacturer, in the first step, sample, assay diluent and anticyclosporine-antibody-coated paramagnetic particles are combined. Cyclosporine present in the sample binds to the antibody-coated microparticles. After incubation and washing, cyclosporine-acridinium-labeled conjugate is added in the second step. Following another incubation and wash, pretrigger (hydrogen peroxide) and trigger solutions (sodium hydroxide) are then added to the reaction mixture. The acridinium undergoes an oxidative reaction when exposed to peroxide and an alkaline solution, producing chemiluminescent reaction thus releasing energy (light emission). The resulting chemiluminescent reaction is measured as relative light units (RLUs). A direct relationship exists between the amount of cyclosporine in the sample and RLU detected by Architect System optics. In one study, the authors concluded that the CMIA Architect has significant reduced cyclosporine metabolite interference relative to other immunoassays and is a convenient and sensitive automated method to measure cyclosporine in whole blood.[16]

CMIA can also be used for analysis of large molecules. The CMIA HIV antigen/antibody combo assay for application on the Architect i system is also a two-step immunoassay to determine the presence of HIV-1 p24 antigen, antibodies to HIV-1 (group M and group O), and antibodies to HIV-2 in human serum or plasma. A relationship exists between the amount of HIV antigen and antibodies in the sample and the RLUs detected by the Architect i System optics. The presence or absence of HIV-1 p24 antigen or HIV-1/HIV-2 antibodies in the specimen is determined by comparing the chemiluminescent signal in the reaction with the cutoff signal determined from the calibration curve. This assay has good clinical performance.[17]

Electrochemiluminescent Immunoassay

An example where the small label is activated electrochemically is the Elecsys automated immunoassay system from Roche Diagnostics, which can be run using cobas and other analyzers available from Roche. For example, in the electrochemiluminescent assay for cyclosporine and tacrolimus the analyte of interest is extracted from whole blood using a pretreatment solution provided by the manufacturer (Roche), and then in the first step the pretreated sample (20 μL) is incubated with the cyclosporine- or tacrolimus-specific biotinylated antibody and ruthenium (Ru)-labeled cyclosporine or tacrolimus derivative for 9 min. In the second step, streptavidin-coated magnetic microparticles are added followed by 9 min incubation. During the second incubation, the entire complex is bound to the solid phase through the interaction between biotin and streptavidin. The reaction mixture after the second incubation is aspirated into the measuring cell and the microparticles are attracted to the electrode by magnetic force, followed by a washing step to remove

any unbound substance. The application of voltage to electrodes induces a chemiluminescent reaction that is measured by a photomultiplier. The analytical performances of the Elecsys cyclosporine and Elecsys tacrolimus assays are acceptable for routine therapeutic drug monitoring of cyclosporine and tacrolimus.[18]

Antibody-Conjugated Magnetic Immunoassays

Antibody-conjugated magnetic immunoassays (ACMIAs) marketed by Siemens include CSA Flex (Cyclosporine), CSA-E Flex reagent cartridges (Cyclosporine Extended Range), TAC-R Flex reagent cartridges (Tacrolimus), and ACMIA sirolimus assay for application on the Dimension systems. The ACMIAs for immunosuppressants are the only assays that do not require any specimen pretreatment. For example, for measuring cyclosporine concentration in the whole blood using ACMIA cyclosporine assay, the whole blood specimen is directly added to a sample cup followed by online mixing and ultrasonic lysing of whole blood to release cyclosporine. An anticyclosporine antibody/β-galactosidase conjugate is then added to the reaction mixture, which binds cyclosporine in the sample. Unbound (excess) conjugate is removed magnetically by adding cyclosporine-coated magnetic beads. The supernatant containing the cyclosporine-antibody-enzyme complex is transferred to a measuring cuvette where β-galactosidase reacts with a chromogenic substrate producing chlorophenol red. The change in absorption at 577 and 700 nm is directly proportional to the cyclosporine concentration in whole blood. ACMIA results are accurate and reliable for monitoring of cyclosporine and tacrolimus in pediatric transplant recipients.[19] However, falsely elevated cyclosporine and tacrolimus levels due to interference of endogenous antibodies, such as rheumatoid factors or heterophilic antibodies, have been reported in ACMIA cyclosporine and tacrolimus assays.[20,21]

Enzyme-Linked Immunosorbent Assay

The label is an enzyme in the enzyme-linked immunosorbent assay (ELISA) that generates chemiluminescent, fluorometric, or colorimetric signal depending on the enzyme substrates used. Examples of commercial automated assay systems using the ELISA technology and chemiluminescent labels are Immulite from Siemens and ACCESS from Beckman Coulter.[22,23] Another type of heterogeneous immunoassay uses polystyrene particles. If these are particles are micro sized, then that type of assay is called microparticle enzyme immunoassay.

SPECIMEN TYPES

Serum and plasma are the most common types of specimen used in immunoassays. Whole-blood specimens must be used for some analytes, such as the immunosuppressant drugs (cyclosporine, tacrolimus, sirolimus, and everolimus), although another immunosuppressant drug, mycophenolic acid, can be monitored in serum or plasma. Urine is the most commonly used specimen in drugs of abuse testing. Urine samples are less frequently affected by hemoglobin or icterus. Turbidity interference is possible in urine, but the cause is most likely bacterial growth or urate precipitation. Preservatives in urine, such as acetic acid, boric acid, or alkali, may interfere in some urine assays. Cerebrospinal fluid (CSF) specimens are used in monitoring the integrity of the blood-brain barrier (by analyzing plasma proteins) or infection in CSF. The most common CSF-interfering substance is blood contamination with hemolysis. Interference from turbidity is also possible in such specimens. Other types of specimens used for immunoassays are saliva, sweat, tears, ascitic and stomach fluids, and bronchial secretions. Hair and nail specimens have been used to provide evidence of a long-term history of drug abuse.[24,25] Amniotic fluid, cord blood, meconium, and breast milk have been used to determine fetal and perinatal exposure to drugs.[26-28] Immunoassays are commercially available for a variety of analytes, both small molecules and large molecules (Table 1.2).

TABLE 1.2
Application of Immunoassays in Clinical Laboratory for the Analysis of Various Analytes

Class of Analyte	Examples of Specific Analyte
Alcohol biomarkers	Carbohydrate-deficient transferrin, β-hexosaminidase, ethyl glucuronide, etc.
Anemia markers	Ferritin, folate, vitamin B_{12}, etc.
Autoantibodies to diagnose and monitor autoimmune diseases	Antinuclear antibody, rheumatoid factor, antithyroid autoantibodies (anti-thyroperoxidase antibodies, anti-thyroglobulin antibodies, anti–thyrotropin receptor antibodies), etc.
Cardiac markers	Cardiac troponin I, troponin T, NT-proBNP, BNP, myoglobin, etc.
Diabetes markers	Insulin, C-peptide, hemoglobin A_{1c}
Therapeutic drug monitoring	Various drugs but most commonly digoxin, procainamide, lidocaine, phenytoin, carbamazepine, phenobarbital, valproic acid, lamotrigine, primidone, theophylline, vancomycin, aminoglycosides, methotrexate, tricyclic antidepressants, cyclosporine, tacrolimus, sirolimus, everolimus, mycophenolic acid, etc.
Drugs of abuse (screening only)	Amphetamine, 3,4-methylenedioxmethamphetamine, barbiturates, benzodiazepines, cocaine metabolite (benzoylecgonine), opiates, oxycodone, hydrocodone, methadone, propoxyphene, phencyclidine, marijuana metabolite (11-nor-delta9-tetrahydrocannabinol-9-carboxylic acid, THC-COOH), lysergic acid diethylamide, etc.
Hormones: reproductive, gastrointestinal, metabolic, etc.	Adrenocorticotropin, growth hormone, follicle-stimulating hormone, luteinizing hormone, prolactin, parathyroid hormone, insulin, cortisol, etc.
Liver disease markers	α2-Macroglobulin, hyaluronic acid, galectin-3, etc.
Thyroid markers	T_4, FT_4, T_3, FT_3, thyroid-stimulating hormone, etc.
Infectious diseases	Various antigen and antibody for diagnosis of different infectious diseases including hepatitis and HIV.
Tumor markers	Prostate-specific antigen, cancer antigens (CA-125, CA-19-9, etc.), β2-microglobulin, α-fetoprotein, carcinoembryonic antigen, human chorionic gonadotropin, etc.

FT_3, free triiodothyronine; FT_4, free thyroxine; *HIV*, human immunodeficiency virus; *NT-proBNP*, N-terminal prohormone brain natriuretic peptide.

ISSUES OF INTERFERENCE

Although immunoassays are widely used in clinical laboratories, interferences in these assays causing both false-positive and false-negative results are serious drawbacks. Both endogenous (bilirubin, hemoglobin, protein, or lipid) and exogenous compounds (drug metabolite, structurally related compounds to target analyte) can cross-react with assay antibody altering immunoassay results, which may cause confusion during diagnosis. Although monoclonal antibody–based immunoassays are less affected by interfering substances than polyclonal antibody–based immunoassays, significant interferences in monoclonal antibody–based immunoassays have also been reported. Moreover, endogenous human antibodies such as heterophilic antibodies or various human antianimal antibodies present in the specimen may also cause interference more commonly with sandwich type immunoassays. In addition, system- or method-related errors, for example, pipetting probe contamination and carryover, may also significantly affect immunoassay results. Various factors that interfere with immunoassays are summarized in Table 1.3.

TABLE 1.3
Endogenous and Exogenous Factors That Interfere With Immunoassays

Interfering Substance	Source	Comments
Bilirubin, hemoglobin, lipids	Endogenous	The components only at highly elevated concentrations (for example, bilirubin >20 mg/dL) interfere with immunoassays. However, most analyzers can also detect these components if present in high concentration in specimen and flag the result, if needed to alter, to the technologist.
Paraproteins	Endogenous	Patients with multiple myeloma often have high concentration of paraproteins. There are reports of both positive and negative interferences of IgM in various assays (therapeutic drugs, thyroxine, etc.).
Heterophilic antibodies	Endogenous	Heterophilic antibodies falsely elevate analyte values most commonly in the sandwich format.
Rheumatoid factors	Endogenous	Rheumatoid factors can falsely increase analyte values most commonly in the sandwich format.
Drug metabolites	Formed endogenously after drug administration	Drug metabolites usually falsely elevate the concentration of parent drug because of the cross-reactivity with the assay antibody. Most serious interferences are reported in falsely elevated immunosuppressant drug levels due to interference of metabolites.
Structurally related compounds	Exogenous	These are the most common cause of assay interference in analysis of small molecules using competitive immunoassays. A common example is false-positive results in amphetamine screening using immunoassay because of the presence of ephedrine/pseudoephedrine.
Biotin	Exogenous	Biotin present in diet or multivitamins (30 μg) has no effect on biotin-based immunoassays. In general, taking biotin supplement up to 5 mg/day may have no effect, but higher dosages or a mega dose (100–300 mg/day) used for treating symptoms of multiple sclerosis affect all immunoassays that use biotin in the assay design. The interference is positive in the analysis of small molecules (competitive format) but negative in the analysis of large molecules (sandwich format).

Interference From High Bilirubin, Hemoglobin, and Lipids

High bilirubin, plasma free hemoglobin, and lipids may cause significant interference with immunoassays by affecting assay signal. In normal adults, total bilirubin concentrations vary from 0.3 to 1.2 mg/dL in serum or plasma. In different forms of jaundice, total bilirubin concentration may increase to as high as 20 mg/dL. Usually total bilirubin concentration below 20 mg/dL does not cause interferences but concentration over 20 mg/dL may cause problem. This is due to absorbance of bilirubin at 454 and 461 nm. Although bilirubin may also interfere with an assay by chemically reacting with a component of the reagent, this is rarely observed.

Hemoglobin is mainly released from hemolysis of red blood cells. Hemolysis can occur in vivo, during venipuncture and blood collection, or during processing of the sample. Hemoglobin interference depends on its concentration in the sample. Serum appears hemolyzed when the serum or plasma free hemoglobin concentration exceeds 20 mg/dL. The absorbance maxima of the heme moiety in hemoglobin are at 540–580 nm wavelengths. However, hemoglobin begins to absorb at around 340 nm and then absorbance increases at 400–430 nm. Interference by hemoglobin (if the specimen is grossly hemolyzed) is due to interfering with optical detection system of the assay.

All lipids in plasma exist as complexed with proteins that are called lipoproteins and the particle size varies from 10 to 1000 nm (the higher the percentage of the lipid, the lower the density of the resulting lipoprotein and the larger the particle size). The lipoprotein particles with high lipid contents are micellar and are the main source of assay interference. Unlike bilirubin and hemoglobin, lipids normally do not participate in chemical reactions and mostly cause interference in assays due to their turbidity and capability of scattering light as in nephelometric assays.

Hasanato et al. studied effects of high bilirubin concentration, hemolysis, and lipemia on various immunoassays using the Abbott Architect analyzer and observed that serum ferritin and TSH levels were overestimated due to hemolysis. In contrast, vitamin B_{12} level progressively decreased as the amount of hemolysis increased. Moreover, significant decrease in progesterone concentration due to lipemia was also observed. For icteric interferences, a strong inverse correlation was observed for folic acid.[29] Steen et al.[30] studied the influence of interference by hemolysis, icterus, and lipemia on the results of 32 analytes in serum or plasma using the Beckman LX-20 instrument (Beckman Coulter)

and reported that on the basis of clinical significance, interference due to hemolysis, icterus, or lipemia was present in only 5, 6, and 12 of the analytes studied, respectively.

In another study, the authors investigated the effects of high bilirubin concentrations, hemolysis, and lipemia on various analytes measured on the Roche cobas 6000 analyzer. The authors reported that for most of the chemistry assays, their data were in good agreement with the Roche package inserts. However, some assays had significant interference at lower index values, whereas others were affected at an index higher than the Roche package inserts indicated. In addition, the authors observed the positive interference by hemolysis on alanine aminotransferase, lipase, total protein, potassium, and iron assay, while negative interference was noted on calcium and creatinine kinase assay. Although most of the immunoassays studied were not affected by hemoglobin, bilirubin, and lipids, several therapeutic drugs were affected either positively or negatively by hemolysis, icterus, or lipemia to a certain extent.[31]

LipoClear, a commercially available polymer, can be used as a lipid-clearing agent. Agarwal et al. measured the hemolysis (H), lipemia (T), and icterus (I) indices on the Vitros 5600 analyzer and also treated analytes affected by lipemia with LipoClear followed by reanalysis. The authors reported that all the measured analytes were affected by gross hemolysis (H-index >1000). Low estradiol levels were affected by severe icterus (I-index >20.0). In addition, complement components (C3, C4), ceruloplasmin, haptoglobin, Ig, and vitamin D were significantly affected by moderate (T-index >100) and severe (T-index >500) lipemia. LipoClear treatment significantly attenuated the lipemic interference in these analytes, except for C3, C4, and IgG.[32]

Interference From Other Endogenous Factors

Autoantibodies are endogenous human antibodies that can interfere with a number of analytes including thyroid hormone, prolactin, insulin. Heterophilic antibodies are also human antibodies that interact with assay antibodies, causing false-positive or false-negative results. However, therapeutic antibodies may also interfere with various immunoassays. Macroanalytes or macroenzymes, which are oligomeric or polymeric conjugates of an analyte and/or conjugated with endogenous antibody and rheumatoid factors, may also cause significant interference in immunoassays. Please see Chapter 5 for more detail.

Heterophilic antibodies may be produced in a person in response to exposure to certain animals or animal products, due to infection by bacterial or viral agents, or nonspecifically.

Although many of the Ig clones in normal human serum may display antianimal antibody properties, only those antibodies with sufficient titer and affinity toward the reagent antibody used in the assay may cause clinically significant interference. Among the antianimal antibodies, the most common are human antimouse antibodies because of the widespread use of murine monoclonal antibody products in therapy or imaging. Heterophilic antibody and antianimal antibody interferences are often grouped together as heterophilic antibody interferences. Such interferences have been mostly found with immunometric sandwich assays, but less often with competition assays. Sample dilution and depletion or removal of interfering antibodies has been recommended to remove heterophilic antibody interference. A patient history of exposure to animals or animal products, or autoimmune diseases, alerts the laboratory professionals about the possibility of encountering heterophilic antibody interference in an assay.

As heterophilic antibodies are found mainly in serum, plasma, or whole blood, but not in urine, such interference is absent in the analysis of urine specimen for the same analyte. This gives an excellent way to detect the interference for analytes, which may be present in both matrices. For example, many case studies with false-positive results for human chorionic gonadotropin (hCG) levels in serum/plasma have been described in the literature. In such cases, if β-hCG levels were measured in parallel urine samples, the false results and the resulting dire consequences could have been easily avoided.[33,34]

Heterophilic antibody interference may cause critical impact and clinical misjudgment, resulting in unnecessary follow-up testing and unneeded but potentially dangerous therapy, leading to significant patient morbidity, especially when due to a false-positive hCG (also a cancer marker) measurement in serum without investigating a parallel urine specimen. The fact that such interferences may not be suspected always from patient history, or that such an effect may be transient in nature, complicates the responsibility of the clinical laboratory in reporting accurate results.

Issues of Cross-Reactivity

Cross-reactivity is the most common cause of interference in immunoassays and is caused by compounds with structural resemblance with the target analyte because such compounds may carry similar or the same type of epitopes to the target analyte. Cross-reactivity with the assay antibody causing false-positive (most common) or false-negative results usually affect competitive immunoassays. Therefore, assay interferences due to the presence of compounds that cross-react with assay antibody are commonly observed in therapeutic drug monitoring and drugs of abuse testing. In therapeutic drug monitoring the cross-reactant may be an endogenous substance such as interference of digoxin-like immunoreactive substances in serum digoxin measurement or a drug metabolite. Significant interferences of cyclosporine and tacrolimus metabolites in therapeutic drug monitoring of cyclosporine and tacrolimus have been well documented in the literature. Drugs of abuse testing in urine using immunoassays are also subjected to interferences due to the presence of cross-reactive substances in urine. False-positive amphetamine/methamphetamine immunoassay results due to use of over-the-counter cold medications containing ephedrine/pseudoephedrine has been reported. This is due to structural similarity of ephedrine/pseudoephedrine with methamphetamine. Please see Chapter 6 for issues of interference in therapeutic drug monitoring. Interferences in drugs of abuse testing using immunoassays are discussed in Chapter 7.

BIOTIN INTERFERENCE

Biotin (vitamin B_7 or vitamin H), a water-soluble vitamin, is a member of vitamin B complex that acts as a cofactor for five carboxylase enzymes essential for life because these enzymes play crucial roles in fatty acid synthesis, gluconeogenesis, and amino acid metabolism. Covalent binding of biotin with these enzymes is essential for their enzymatic activities. The recommended daily intake of biotin for adults is only 30 μg. This requirement can be easily satisfied by eating a balanced diet every day. However, people take biotin supplements most commonly for healthy hair, nail, and skin. Biotin supplements are freely available in health food stores in various doses ranging from 500 μg to 100 mg, although in the United States a daily dosage of 5 mg is promoted for healthy hair, nail, and skin. Nutritional biotin deficiency is rare in developed countries but inborn errors of metabolism may cause biotin deficiency. Although oral biotin supplement results in healthy hair and nail growth in individuals with biotin deficiency, currently there is no scientific proof that biotin supplement helps hair growth in healthy individuals without any biotin deficiency. Studies have shown that a daily biotin supplement of 2.5 mg may help cure brittle nail problem. However, people with biotin deficiency may require 10–40 mg biotin per day. For some individuals, higher doses may be needed (40–100 mg) to treat biotin deficiency. Pharmaceutic doses of biotin (100–300 mg) may be useful in treating symptoms of multiple sclerosis as indicated by preliminary results from some clinical trials. Please see Chapters 2 and 3 for a detailed discussion on biotin.

Biotin is not a cross-reactant and has a unique mechanism of interference in immunoassays that utilize biotin in assay design. Eating a balanced meal, taking multivitamin formulations (usually contain 30 μg vitamin), or taking low-dose biotin supplements (500 μg to <5 mg) every day has no effect on immunoassays that utilize biotin in assay design. Significant biotin interference can be observed in people taking 5 mg or more biotin per day. However, the threshold of biotin interference varies widely between various manufacturers. Usually the serum biotin concentration is very low but when it reaches 500 ng/mL, virtually all biotin-based assays are affected. Such high serum biotin level is expected in patients taking pharmacologic dosage of biotin (100–300 mg per day). Please see Chapter 4 for an in-depth discussion on this topic.

Mechanism of Biotin Interference

Biotin only affects immunoassays that utilize biotin in the assay design. The most common use of biotin in assay design is biotinylated antibody. The biotin present in biotinylated antibody has a strong affinity for avidin, a glycoprotein (molecular weight 66 kDa) present in egg white as well as streptavidin, a bacterial protein (molecular weight 60 kDa). Both avidin and streptavidin molecules have four binding sites for biotin, and these biotin-binding proteins can be easily immobilized onto a surface, microparticles, or chromatographic support without compromising the biotin-binding capability. In biotin-based immunoassays the antibody is usually biotinylated, while the antigen–biotinylated antibody complex is captured using streptavidin-coated microparticles.

Biotinylation of antibody is a straightforward procedure that can be achieved by using a succinimidyl ester of biotin. This water-soluble reagent reacts with primary amines of the lysine residues or the amino terminus of the antibody to form amide bond.[35] Biotinylation of antibody does not alter the antigen-binding capacity of the original antibody. Although binding of biotin to streptavidin (or avidin) is noncovalent in nature, because of the strong affinity of streptavidin or avidin for biotin, the complex formed is very stable

and is resistant to breakdown even in the environment of changing pH and temperature or in the presence of denaturants or detergents. Therefore, a chemical reaction or condition that generates assay signal has no effect on biotin-streptavidin interaction.

Biotin interference causes false elevated analyte values using the competitive assay format but false lower values using sandwich format. In a competitive immunoassay using biotinylated antibody, the assay signal is inversely proportional to the analyte concentration. However, in a sandwich immunoassay, the assay signal is directly proportional to the analyte concentration. This difference explains why biotin causes positive interferences in competitive immunoassay format but negative interference in sandwich format.

Competitive immunoassays are used for small molecules with molecular weight less than 1000 Da as well as for molecules with molecular weight higher than 1000 Da, for example, cyclosporine (molecular weight 1202.6 Da), vancomycin (molecular weight 1485.7 Da). In competitive immunoassay the analyte molecules compete with the labeled analyte (the source of assay signal) molecules for limited binding sites on the biotinylated antibody specific for the analyte. After incubation, biotinylated antibodies (bound to either the analyte molecule or the labeled analyte molecule) are captured on a streptavidin-coated solid phase such as a microparticle or magnetic microparticle. In the absence of any interfering substance, if the analyte concentration is high, then few labeled antigen molecules will bind with biotinylated antibody and eventually to the streptavidin-coated solid phase. After washing, signal is generated by the labeled antigen, which should be lower in intensity if the analyte concentration is high. As a result, there is an inverse relationship between analyte concentration in the specimen and the intensity of the signal. If excess biotin is present in the specimen then biotin binds to streptavidin-coated solid phase, thus preventing binding of antibody-labeled analyte (as well as antibody-antigen complexes) with streptavidin. Any antibody that is unbound to the solid phase is removed during washing, and if a high level of biotin is present in the specimen then antibodies bound to labeled antigens producing a signal indicative of the true concentration of the analyte are also washed off during the washing step. Because biotin occupies the binding sites on the streptavidin-coated solid phase, which could otherwise bind biotinylated antibody-antigen complexes, the intensity of the assay signal is reduced and the analyte value calculated from such signal is falsely elevated (Fig. 1.1).

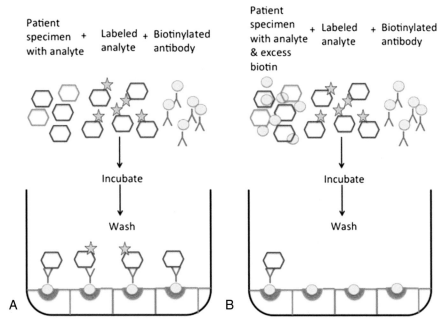

FIG. 1.1 Mechanism of biotin interference in competitive immunoassay format where patient specimen, labeled analyte, and biotinylated antibody are added to a reaction vessel with a streptavidin-coated well. **(A)** The biotinylated antibody will bind to streptavidin, which is attached to the solid phase. The analyte of interest will compete with the labeled analyte for binding to the biotinylated antibodies. Residual specimen and unbound analyte (labeled analyte or the analyte present in the specimen) will be washed off. The remaining signal will be inversely proportional to the concentration of analyte. **(B)** However, if there is excess biotin in the specimen the biotin will bind to the streptavidin sites, blocking the biotinylated antibody and hence the analyte of interest. The biotinylated antibody will bind to the analyte of interest, but without being tethered to the solid phase, it will be washed off, resulting in a signal that is falsely decreased, which will lead to a false increased result. (Adapted from Paula Jenkins C, Greene DN. Biotin interference in clinical immunoassays. *J App Lab Med.* 2018;2(6):941–951. https://doi.org/10.1373/jalm.2017.024257. Reproduced with permission from the American Association for Clinical Chemistry.)

However, in sandwich immunoassay the analyte is sandwiched between the signal antibody and the biotinylated antibody, which links the antibody-analyte sandwich to a streptavidin-coated solid phase. As a result, if no interference is present then the assay signal is directly proportional to the analyte concentration. However, when excess biotin is present in the specimen the biotin molecules will saturate the streptavidin-binding sites, thus preventing antibody-analyte sandwich to bind with the streptavidin-coated solid phase to generate assay signal after the washing phase. Hence, in the presence of high biotin concentrations the intensity of the assay signal would be low, which results in underestimation of the true analyte concentration, thus causing negative interference.[36] This mechanism of negative interference of biotin in sandwich type immunoassay is shown in Fig. 1.2.

CONCLUSIONS

Immunoassays are widely used in clinical laboratories for analysis of many analytes. Usually the competitive assay format is used for analysis of relatively small molecules, while the sandwich assay format (also known as noncompetitive or immunometric format) is used for analysis of large molecules such as proteins. Biotin interference impacts only those immunoassays that utilize biotin in assay design. Eating a balanced diet, taking multivitamins (containing approximately 30 μg of biotin), or taking biotin supplement in small doses have no impact on immunoassays using biotinylated antibody. However, taking 5 mg biotin supplement per day may impact some biotin-based assays, while taking pharmaceutical dosage (100–300 mg/day) for treating symptoms of multiple myeloma or other diseases may impact all biotin-based immunoassays.[37]

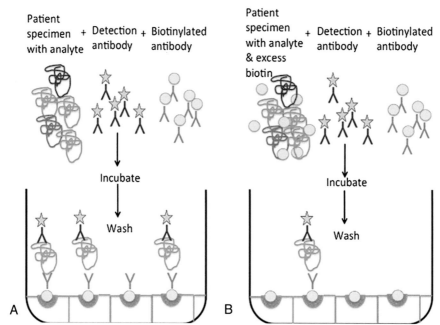

FIG. 1.2 Mechanism of biotin interference in sandwich (noncompetitive) format where the patient specimen, detection antibodies, and biotinylated antibodies are added to a reaction vessel with a streptavidin-coated well. **(A)** The biotinylated antibody will bind to streptavidin, which tethers it to the solid phase. The analyte of interest will be sandwiched between the biotinylated and detection antibodies. Residual specimen and detection antibodies will be washed off. The remaining signal will be directly proportional to the concentration of the analyte. **(B)** However, if there is excess biotin in the specimen the biotin will bind to the streptavidin sites, blocking the biotinylated antibody and hence the analyte of interest. The detection antibody will bind to the analyte of interest, but without being tethered to the solid phase, it will be washed off, resulting in a signal that is falsely decreased, which will lead to a false decreased result. (Adapted from Paula Jenkins C, Greene DN. Biotin interference in clinical immunoassays. *J App Lab Med.* 2018;2(6):941–951. https://doi.org/10.1373/jalm.2017.024257. Reproduced with permission from the American Association for Clinical Chemistry.)

REFERENCES

1. Selby C. Interference in immunoassays. *Ann Clin Biochem.* 1999;36:704–721.
2. Pei X, Zhang B, Tang J, Liu B, et al. Sandwich-type immuno sensors and immunoassays exploiting nanostructure labels: a review. *Anal Chim Acta.* 2013;758:1–18.
3. Köhler G, Milstein C. Derivation of specific antibody-producing tissue culture and tumor lines by cell fusion. *Eur J Immunol.* 1976;6:511–519.
4. Sevier ED, David GS, Martinis J, Desmond WJ, et al. Monoclonal antibodies in clinical immunology. *Clin Chem.* 1981;27:1797–1806.
5. Yalow RS, Berson SA. Immunoassay of endogenous plasma insulin in man. *J Clin Invest.* 1960;39:1157–1175.
6. https://www.mordorintelligence.com/industry-reports/global-immunoassay-technologies-market-industry.
7. Jolley ME, Stroupe SD, Schwenzer KS, et al. Fluorescence polarization immunoassay III. An automated system for therapeutic drug determination. *Clin Chem.* 1981;27:1575–1579.
8. Smith DS, Eremin SA. Fluorescence polarization immunoassays and related methods for simple, high-throughput screening of small molecules. *Anal Bioanal Chem.* 2008;391:1499–1507.
9. Curtis EG, Patel JA. Enzyme multiplied immunoassay technique: a review. *CRC Crit Rev Clin Lab Sci.* 1978;9:303–320.
10. Jeon SI, Yang X, Andrade JD. Modeling of homogeneous cloned enzyme donor immunoassay. *Anal Biochem.* 2004;333:136–147.
11. Kelley WE, Lockwood CM, Cervelli DR, Sterner J, et al. Cardiovascular disease testing on the Dimension Vista® system: biomarkers of acute coronary syndromes. *Clin Biochem.* 2009;42:1444–1451.

12. Monneret D, Guergour D, Vergnaud S, Laporte F, et al. Evaluation of LOCI technology-based thyroid blood tests on the Dimension Vista analyzer. *Clin Biochem.* 2013;46:1290–1297.

13. Ullman EF, Kirakossian H, Switchenko AC, Ishkanian J, et al. Luminescent oxygen channeling assay (LOCI): sensitive, broadly applicable homogeneous immunoassay method. *Clin Chem.* 1996;42:1518–1826.

14. Patel R, Pollner R, de Keczer S, Pease J, et al. Quantification of DNA using the luminescent oxygen channeling assay. *Clin Chem.* 2000;46:1471–1477.

15. Dai JL, Sokoll LJ, Chan DW. Automated chemiluminescent immunoassay analyzers. *J Clin Ligand Assay.* 1998;21:377–385.

16. Serdarevic N, Zunic L. Comparison of architect I 2000 for determination of cyclosporine with AxSYM. *Acta Inf Med.* 2012;20:214–217.

17. Cui C, Liu P, Feng Z, Xin R, et al. Evaluation of the clinical effectiveness of HIV antigen/antibody screening using a chemiluminescence microparticle immunoassay. *J Virol Methods.* 2015;214:33–36.

18. Sasano M, Kimura S, Maeda I, Hidaka Y. Analytical performance evaluation of the Elecsys® Cyclosporine and Elecsys® Tacrolimus assays on the Cobas e411 analyzer. *Pract Lab Med.* 2017;8:10–17.

19. Cangemi G, Barco S, Bonifazio P, Maffia A, et al. Comparison of antibody-conjugated magnetic immunoassay and liquid chromatography-tandem mass spectrometry for the measurement of cyclosporine and tacrolimus in whole blood. *Int J Immunopathol Pharmacol.* 2013;26:419–426.

20. Morelle J, Wallemacq P, Van Caeneghem O, Goffin E. Clinically unexpected cyclosporine levels using the AC-MIA method on the RXL dimension analyser. *Nephrol Dial Transplant.* 2011;26. 1428–1231.

21. D'Alessandro M, Mariani P, Mennini G, Severi D, et al. Falsely elevated tacrolimus concentrations measured using the ACMIA method due to circulating endogenous antibodies in a kidney transplant recipient. *Clin Chim Acta.* 2011;412:245–248.

22. Babson AL, Olsen DR, Palmieri T, et al. The IMMULITE assay tube: a new approach to heterogeneous ligand assay. *Clin Chem.* 1991;37:1521–1522.

23. Christenson RH, Apple FS, Morgan DL. Cardiac troponin I measurement with the ACCESS® immunoassay system: analytical and clinical performance characteristics. *Clin Chem.* 1998;44:52–60.

24. Pichini S, Pacifici R, Altieri I, Pellegrini M, Zuccaro P. Determination of opiates and cocaine in hair as trimethylsilyl derivatives using gas chromatography-tandem mass spectrometry. *J Anal Toxicol.* 1999;5:343–348.

25. Palmeri A, Pichini S, Pacifici R, Zuccaro P, Lopez A. Drugs in nails: physiology, pharmacokinetics and forensic toxicology. *Clin Pharmacokinet.* 2000;38:95–110.

26. Pichini S, Altieri I, Zuccaro P, Pacifici R. Drug monitoring in nonconventional biological fluids and matrices. *Clin Pharmacokinet.* 1996;30:211–228.

27. Wood KE, Krasowski MD, Strathmann FG, McMillin GA. Meconium drug testing in multiple births in the USA. *J Anal Toxicol.* 2014;38:397–403.

28. Dickson PH, Lind A, Studts P, Nipper HC, Makoid M, et al. The routine Analysis of breast milk for drugs of abuse in a clinical toxicology laboratory. *J Forensic Sci.* 1994;39:2341–2345.

29. Hasanato R, Brearton S, Alshebani M, Bailey L, et al. Effects of serum indices interference on hormonal results from the Abbott Architect i2000 immunoassay analyser. *Br J Biomed Sci.* 2015;72:151–155.

30. Steen G, Vermeer HJ, Naus AJ, Goevaerts B, et al. Multi-center evaluation of the interference of hemoglobin, bilirubin and lipids on SYNCHRON LX-20 assays. *Clin Chem Lab Med.* 2006;44:413–419.

31. Ji JZ, Meng QH. Evaluation of the interference of hemoglobin, bilirubin, and lipids on Roche Cobas 6000 assays. *Clin Chim Acta.* 2011;412:1550–1553.

32. Agarwal S, Vargas G, Nordstrom C, Tam E. Effect of interference from hemolysis, icterus and lipemia on routine pediatric clinical chemistry assays. *Clin Chim Acta.* 2015;438:241–245.

33. Albersen A, Kemper-Proper E, Thelen MHM, et al. A case of consistent discrepancies between urine and blood human chorionic gonadotropin measurements. *Clin Chem Lab Med.* 2011;49:1029–1032.

34. Braunstein G. False positive serum human chorionic gonadotropin results: causes, characteristics, and recognition. *Am J Obstet Gynecol.* 2002;187:217–224.

35. Mao SY. Biotinylation of antibodies. *Meth Mol Biol.* 2010;588:49–52.

36. Jenkins Colon P, Greene DN. Biotin interference in clinical immunoassays. *J App Lab Med.* 2018;2:941–951.

37. Piketty ML, Prie D, Sedel F, Bernard D, et al. High-dose biotin therapy leading to false biochemical endocrine profiles: validation of a simple method to overcome biotin interference. *Clin Chem Lab Med.* 2017;55:817–825.

CHAPTER 2

Biotin: Pharmacology, Pathophysiology, and Assessment of Biotin Status

INTRODUCTION

Biotin, known as vitamin B_7 or vitamin H ("H" is the initial for "Haar" and "Haut" the German word for hair and skin) is a member of the vitamin B complex and is water soluble. Biotin was first isolated in 1936 as a yeast growth factor from egg yolk by Kogl and Tonnis, and its structure ($C_{10}H_6O_3N_2S$, molecular weight 244.3 Da) was elucidated in 1942 by du Vigneaud et al. Biotin is a heterocyclic ring that is attached to the aliphatic side chain terminating in a carboxyl group. Out of the eight different stereoisomers, only d-biotin (which is usually referred to as biotin) exhibits biological activity and is found in nature. Biotin is present in most plant and animal tissues and is synthesized by a variety of bacteria.[1] Biotin is essential for normal cellular functions, growth, and development because it acts as a cofactor for five carboxylases (four located in mitochondria and one in cytoplasm). These carboxylases play a critical role in the intermediate metabolism of gluconeogenesis, fatty acid synthesis, and amino acid catabolism.[2] Humans and other mammals cannot synthesize biotin and must obtain biotin from diet via intestinal absorption. In addition, normal microflora of the large intestine can synthesize biotin. Chemical structure of biotin is given in Fig. 2.1.

EGG WHITE INJURY

Seborrheic dermatitis of infancy, a self-limiting skin eruption of infants (<6 months of age), closely resembles skin changes observed in rats fed a diet rich in egg white. Biotin is effective in treating seborrheic dermatitis. Therefore skin change in infants with seborrheic dermatitis is a sign of biotin deficiency. Egg white contains avidin, a glycoprotein (molecular weight 66–60 kDa) that strongly binds biotin.[3] Therefore biotin deficiency due to prolonged consumption of raw eggs is called "egg white injury." Baugh et al.[4] reported a case of biotin deficiency induced by raw egg consumption in a cirrhotic patient. In the stomach and intestinal lumen, avidin tightly binds biotin and thereby prevents intestinal absorption of biotin. Cooking denatures avidin thus destroying its biotin-binding capacity and as a result biotin present in egg yolk and other food sources can be absorbed from the intestine.

Case Report: A 66-year-old man was admitted to the hospital for sudden loss of consciousness, tonic-clonic movement, and psychomotor agitation. During the previous 4 months the patient had complained of drowsiness, fatigue, lack of appetite, and nausea. During the previous 2 months, he also experienced hair loss, coating of the tongue, and skin flaking with pruritus. The social history revealed that during the previous year, the patient established a chicken farm and each morning when he went to the farm to collect newly laid eggs, he ate some raw eggs because their shells were broken. Despite the use of sophisticated technologies, no diagnosis was reached. At that point the authors measured the serum biotin level, which was low (98 pmol/L) thus indicating egg white injury syndrome due to biotin deficiency. The patient was successfully treated with biotin supplement and his symptoms resolved. After 1 week of biotin supplement, his serum biotin level was elevated to 124 pmol/L (normal reference range, 195–300 pmol/L), and after 6 months of supplementation, his serum biotin level was 204 pmol/L. The involvement of biotin in fat synthesis is often cited as a cause of dermatologic problems associated with biotin deficiency, while the second most common clinical presentation of biotin deficiency is neurologic problems. Both glucose and fats are used for energy within the nervous system, and biotin deficiency interferes with glucose and fat synthesis. If untreated, neurologic problems may cause depression and generalized muscular pain (myalgias), hyperesthesias, and paresthesias.[5]

Case Report: A developmentally normal and well-nourished 10-year-old boy was admitted to the hospital for limb weakness. Initially the boy had frequent

Biotin and Other Interferences in Immunoassays. https://doi.org/10.1016/B978-0-12-816429-7.00002-2

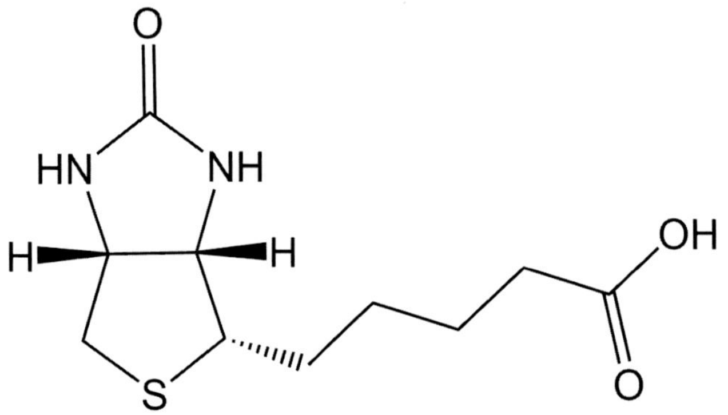

FIG. 2.1 Chemical structure of biotin. (Courtesy of Matthew D. Krasowski, MD, PhD, Vice Chair for Clinical Pathology and Laboratory Services, University of Iowa Carver College of Medicine.)

falls and an unsteady gait due to weakness of the distal muscle of his lower limbs, which later progressed to his proximal and trunk muscles so that he could neither move his limbs nor get up from the bed. He was also anorexic, lethargic, and not talking. A retrospective history revealed that the boy had been eating raw eggs every alternate day for the past 1 year for extra nutrition, which probably caused biotin deficiency. A thorough investigation established diagnosis of biotin-responsive basal ganglia disease (BBGD), and he was successfully treated with oral biotin (5 mg/kg per day).[6]

BIOTIN CONTENT IN FOODS

Biotin is present in appreciable quantities in egg yolks, meat, fish, dairy, and some vegetables. Staggs et al. measured the biotin content of 87 foods using acid hydrolysis to liberate bound biotin followed by high-performance liquid chromatography/avidin-binding assays. In general, foods such as liver, egg yolk, and green vegetables are good sources of biotin. However, the authors commented that their measured values in some foods differed significantly from previously published values because the chromatographic method used by authors to measure biotin in various foods is analytically more robust than the methods used by other authors.[7] Some foods with high biotin content are listed in Table 2.1.

PHARMACOLOGY OF BIOTIN

Biotin is a water-soluble vitamin that is readily absorbed from the intestine. In fact, human intestine is exposed to two sources of biotin: one from the consumed food and the other synthesized by normal microflora of the large intestine. Biotin in food is largely protein bound.

The protein-bound biotin is digested to free form prior to absorption in the small intestine. Digestion is first performed by the action of gastrointestinal proteases and peptidases producing biotin-L-lysine (biocytin) and biotin-short peptides (biotin-oligopeptides), which are then converted into free biotin by the action of the biotinidase enzyme.[8] Absorption of free biotin in the small and large intestines involves a saturable sodium-dependent carrier–mediated process that is shared with pantothenic acid and α-lipoic acid. This transporter system is known as sodium-dependent multivitamin transporter. In humans, this transporter is called the human sodium-dependent multivitamin transporter (hSMVT). The hSMVT protein (635 amino acids) is produced by the *SLC5A6* gene, which is located in chromosome 2p23 and consists of 17 exons.[9] The human intestinal biotin uptake processes are adaptively upregulated by biotin deficiency via a transcriptionally mediated mechanism involving the sites of Kruppel-like factor 4. Bioavailability of dietary biotin varies from 5% to almost 100% depending on the food type.[10] Interestingly, SMVT not only mediates intestinal absorption of biotin but also is crucial for biotin uptake into the liver and peripheral tissues and for renal reabsorption.[11] In addition, monocarboxylate transporter 1 might account for biotin uptake in some cell lineages such as lymphoid cells and keratinocytes.[12]

Interestingly, biotin has almost 100% bioavailability if given as a supplement and also in pharmaceutical doses.[13] However, large doses of pantothenic acid (vitamin B$_5$) has the potential to compete with biotin for intestinal and cellular uptake by hSMVT.[14]

After absorption, biotin is distributed in the serum and tissues. Based on plasma analysis of 11 normal adults, Mock and Malik[15] reported that approximately 12% of total biotin in plasma is covalently bound to

TABLE 2.1 Foods That Are Good Sources of Biotin	
Food	**Biotin per Serving (μg)**
Cooked beef liver	30.8
Cooked chicken liver	138.0
Cooked whole egg	10.0
Cooked egg white	2.02
Cooked hamburger patty	1.65
Hot dog and cooked chicken	2.06
Cooked pork chop	3.57
Tuna canned in water	0.43
Pink salmon canned in water	3.69
2% Milk	0.27
American cheese	0.59
Cheddar cheese	0.40
Plain yogurt	0.14
Fresh broccoli	1.07
Canned mushroom	2.59
Cooked sweet potato	1.16
Fresh tomato	0.30
Fresh strawberry	1.67
Banana pudding	1.73
Chili	2.29
Roasted and salted almonds	1.32
Roasted and salted peanuts	4.91
Roasted and salted sunflower seeds	2.42

Source of data, reference Staggs CG, Sealey WM, McCabe BJ, Teague AM, et al. Determination of the biotin content of select foods using accurate and sensitive HPLC/avidin binding. *J Food Compost Anal* 2004;17:767–776.

TABLE 2.2 Catabolism of Biotin	
Catabolic Pathway	**Metabolites Formed**
β-Oxidation	Bisnorbiotin, Tetranorbiotin, bisnorbiotin methyl ketone, tetranorbiotin methyl ketone
Sulfur oxidation	Biotin sulfoxide, biotin sulfone
Both β-oxidation and sulfur oxidation	Bisnorbiotin sulfone

β-ketobiotin and β-ketobisnorbiotin are unstable and may undergo decarboxylation yielding bisnorbiotin methyl ketone and tetranorbiotin methyl ketone, respectively. Whether β-oxidation of biotin occurs in mitochondria or peroxisomes is unclear. Oxidation of sulfur in the heterocyclic ring of biotin produces biotin sulfoxide and biotin sulfone. Most likely, the sulfur oxidation of biotin takes place in the smooth endoplasmic reticulum by the nicotinamide adenine dinucleotide phosphate (NADPH)-dependent pathway.[16] Biotin is also catabolized by a combination of both β-oxidation and sulfur oxidation yielding metabolites such as bisnorbiotin sulfone. Catabolic pathways of biotin are summarized in Table 2.2.

In general, biotin undergoes limited degree of metabolism in humans, as more than half of the ingested biotin is recovered unchanged in urine.[17] When biotin was administered orally in normal subjects in four different doses in six healthy volunteers, for all four doses, unchanged biotin accounted for >50% of the total biotin and metabolites in urine, bisnorbiotin 13%–23%, biotin sulfoxide 5%–13%, bisnorbiotin methyl ketone 3%–9%, and biotin sulfone (1%–3%).[18]

Although the half-life of biotin is approximately 1.8 h after taking microgram amounts, after an intake of 300 mg biotin, half-life may be prolonged to 7.8–18.8 h .[19] Usually normal serum concentration of biotin is very low in individuals taking no biotin supplement. Biotin deficiency only occurs at a serum level of <0.2 ng/mL. Suboptimal level is 0.2–0.4 ng/mL, while biotin level above 0.4 ng/mL is considered adequate. For subjects not taking supplement, biotin concentrations usually vary from 0.4 to 1.2 ng/mL.[20] However, serum biotin levels are significantly higher in individuals taking biotin supplements or pharmaceutical doses of biotin. Serum biotin levels may approach 1200 ng/mL in patients taking 300 mg biotin supplement per day. Usually maximum biotin level in serum is observed 2 h after oral intake of biotin. Please see

serum protein, 7% is reversibly bound to serum protein, and 81% exists in the free form. Circulating biotin undergoes filtration in the renal glomeruli and is then reabsorbed by renal proximal tubular epithelial cells.

Biotin Catabolism and Half-Life

McCormick et al. elucidated biotin catabolism showing two regions of the biotin molecule that are primarily catabolized. Biotin is catabolized by β-oxidation at the valeric acid side chain, and repeated cleavage of two carbon units leads to the formation of bisnorbiotin, tetranorbiotin, and related metabolites. However,

TABLE 2.3
Various Parameters of Biotin

Parameter	Comments
Water solubility	Soluble in water.
Absorption	Protein-bound biotin is first converted into free biotin by enzymes in the intestine. Then free biotin is absorbed from small and large intestines where human sodium-dependent multivitamin transporter plays an important role.
Bioavailability	5% to almost 100% from food sources. Almost 100% from supplements.
Major metabolites	Unchanged biotin accounted for >50% of the total biotin. Major metabolites in urine are bisnorbiotin, 13%–23%; biotin sulfoxide, 5%–13%; bisnorbiotin methyl ketone, 3%–9%; and biotin sulfone, 1%–3%.
Half-life	Approximately 1.8 h after taking microgram amounts, but after intake of 300 mg biotin, half-life may be prolonged to 7.8–18.8 h.
Serum levels	For subjects not taking supplement, biotin concentrations in serum usually vary from 0.4 to 1.2 ng/mL. However, serum biotin levels are significantly increased after taking biotin supplements and may reach 1200 ng/mL after taking 300 mg of biotin supplement.
Maximum serum concentration	Usually 1–3 h after taking oral supplement.
Drug interaction	Prolonged treatment with anticonvulsants may cause biotin deficiency.

Chapter 4 for more in-depth discussion on serum biotin levels. The various pharmacokinetic parameters of biotin are given in Table 2.3.

Biotin Toxicity

Biotin is a water-soluble vitamin and is nontoxic. In diabetic people without disorders of biotin metabolism, oral biotin supplementation of 5 mg per day for 2 years did not show any adverse effects of biotin. Interestingly, biotin supplementation was associated with reduced pain of diabetic neuropathy where beneficial effects were observed with 4–8 weeks of initiation of biotin therapy.[21] In one study involving 23 consecutive patients with primary and secondary progressive multiple sclerosis, the authors showed that supplementation of 100–300 mg biotin per day over a period of 2–36 months showed significant clinical improvement in most patients. The authors reported no toxicity of biotin indicating that biotin is safe even at a high pharmaceutical dose.[22] However, there is an isolated case report of toxicity due to concurrent use of biotin and pantothenic acid.

Case Report: A 76-year-old white woman was hospitalized with chest pain and difficulty with breathing. The patient had no history of allergy and at the time of hospitalization was receiving trimetazidine for 6 years. She was also taking biotin (10 mg/day) and pantothenic acid (300 mg/day) for last 2 months for alopecia. Blood test revealed an inflammatory syndrome with erythrocyte sedimentation rate of 51 mm/h and eosinophil count of 1200–1500 cells/μL (normal, 500 or less). Moreover, histologic examination of the pericardial biopsy revealed an eosinophilic infiltrate. There were no nuclear antibodies and no rheumatic factor. Screening for viruses, parasites, bacteria, and malignant tumor were all negative. The patient had life-threatening eosinophilic pleuropericardial effusion related to biotin and pantothenic acid. After withdrawal of vitamins, her symptoms resolved and the eosinophilia disappeared. The authors commented that other drugs known to cause eosinophilic pleuropericarditis are cephalosporins, dantrolene, propylthiouracil, and nitrofurantoin. This is the first case report of eosinophilic pleuropericarditis due to the use of biotin and pantothenic acid. In January 1985 the Food and Drug Administration withdrew its approval of biotin products for treatment of male pattern baldness.[23]

Because of one isolated report of combined biotin and pantothenic acid toxicity, there is no established upper tolerable intake of biotin.

Biotin-Drug Interactions

Although decreased urinary excretion of biotin and its metabolite bisnorbiotin is an early and sensitive indicator of biotin deficiency, decreased serum biotin concentration is not a good indicator of biotin deficiency. Increased urinary excretion of 3-hydroxyisovaleric acid, a leucine metabolite that is excreted in increased quantities with deficiency of the biotin-dependent enzyme β-methylcrotonyl-CoA carboxylase, is an early sensitive marker of biotin deficiency [24] In patients undergoing long-term therapy with carbamazepine and/or phenytoin, reduced plasma concentrations of biotin have been reported. In addition, pathologic organic aciduria has also been reported with long-term anticonvulsant therapy, suggesting biotin deficiency in these patients. In one study based on 7 children receiving carbamazepine and/or phenytoin, 6 children receiving phenobarbital, and 60 healthy children, the authors reported increased excretion of bisnorbiotin in children receiving carbamazepine, phenytoin, or phenobarbital. However, excretion of biotin sulfoxide was increased only in children receiving carbamazepine and or phenytoin. In addition, 3-hydroxyisovaleric acid urinary excretion was increased significantly in six children receiving carbamazepine and/or phenytoin, but in one child, its excretion was decreased. The authors concluded that long-term administration of some anticonvulsants can accelerate biotin catabolism.[25]

Krause et al. microbiologically determined plasma biotin levels in 404 epileptic patients under long-term treatment with anticonvulsants and observed that biotin levels were markedly lower in epileptic patients receiving certain anticonvulsants than those in 112 controls. In contrast, epileptic patients treated with valproate sodium in monotherapy showed considerably higher biotin levels. The authors concluded that long-term treatment of epileptic patients with primidone, carbamazepine, phenytoin, or phenobarbital was associated with biotin deficiency but treatment with valproic acid did not lower serum biotin levels.[26] In another study, the authors also demonstrated that patients receiving valproic acid showed normal serum biotin levels and no change in urinary excretion of organic acid.[27]

Long-term treatment with antibacterial drugs may decrease synthesis of biotin by gut bacteria but the extent to which such biotin contributes to human nutrition is not well characterized. There is no study to show biotin deficiency in humans after prolonged therapy with antibiotics.

Lipoic acid (also known as α-lipoic acid) competes with biotin for binding with SMVT, thus potentially decreasing the cellular uptake of biotin. Chronic administration of pharmacologic doses of lipoic acid decreased the activities of the biotin-dependent enzymes pyruvate carboxylase and 3-methylcrotonyl-CoA carboxylase (3-MCC) in rat liver to 64%–72% of control values.[28]

PATHOPHYSIOLOGY OF BIOTIN

Biotin is an essential vitamin but humans cannot synthesize biotin. Therefore the daily requirement must be fulfilled through diet. Although biotin is synthesized by normal microflora of the large intestine, contribution of biotin from this source is uncertain in biotin status of humans. For biotin, Food and Nutrition Board of the National Research Council recommends 30 µg daily intake for adults and pregnant women.[29] The dietary biotin intake in the Western population has been estimated to be 35–70 µg per day.[30,31]

Biotin-Dependent Enzymes

The major biological function of biotin is to act as a covalently bound cofactor for the biological activities of five mammalian biotin-dependent carboxylases. These biotin-dependent carboxylases have a crucial role in essential biological processes including
- fatty acid synthesis
- gluconeogenesis
- amino acid metabolism

The term "biotinylation" is referred to the covalent attachment of biotin with any molecule including apocarboxylases (catalytically inactive carboxylases). In mammals, holocarboxylase synthetase catalyzes the covalent binding of biotin with the ε-amino group of lysine in different apocarboxylases to form the active enzyme known as holocarboxylases. Biotinylation of carboxylases requires ATP. Five mammalian carboxylases, namely, acetyl-CoA carboxylase 1 (cytosolic form), acetyl-CoA carboxylase 2 (mitochondrial form), pyruvate carboxylase, propionyl-CoA carboxylase, and 3-MCC, require biotin as a cofactor. Biotin-dependent carboxylases use bicarbonate as a CO_2 donor, and in the first step of the reaction, this moiety is attached with the N1 atom of biotin. This step is ATP and magnesium dependent. In the final step of enzymatic reaction, the CO_2 group is transferred from carboxybiotin to the acceptor of the carboxyl group referred to as the "substrate." Most substrates are coenzyme A (CoA) esters of organic acids such as acetyl-CoA, propionyl-CoA, and 3-methylcrotonyl-CoA. In addition, small molecules such as pyruvate can also act as a substrate.[32]

Both acetyl-CoA carboxylase 1 and 2 catalyze the conversion of acetyl-CoA to malonyl-CoA in the presence of ATP and bicarbonate. Malonyl-CoA produced in the cytoplasm by acetyl-CoA carboxylase 1 is essential for the biosynthesis of fatty acids because it provides two-carbon building blocks for fatty acid biosynthesis and subsequently other lipids including triglycerides and phospholipids. Phospholipids are an essential component of cell membrane. The acetyl-CoA carboxylase 2 produces malonyl-CoA from acetyl-CoA in the mitochondria, which acts as a regulator of β-oxidation of fatty acids because malonyl-CoA is a potent inhibitor of carnitine palmitoyltransferase I, the crucial enzyme needed for transport of long-chain fatty acyl-CoA into mitochondria for β-oxidation. Therefore malonyl-CoA produced by acetyl-CoA carboxylase 2 localized in the mitochondria downregulates β-oxidation of fatty acid in mitochondria.

Pyruvate carboxylase localized in mitochondria is a key enzyme in gluconeogenesis and also acts to provide a tricarboxylic acid cycle intermediate. Pyruvate carboxylase catalyzes the ATP-dependent incorporation of bicarbonate to pyruvate producing oxaloacetate. Then oxaloacetate is converted into phosphoenolpyruvate and eventually through a sequence of steps into glucose.

Propionyl-CoA is the end product of catabolism of isoleucine, valine, methionine, threonine, cholesterol side chain, and fatty acids with an odd number of carbon atoms. Propionyl-CoA carboxylase catalyzes conversion of propionyl-CoA into methylmalonyl-CoA, a step important for the catabolism of certain amino acids and odd chain fatty acids. The (S)-methylmalonyl-CoA initially formed is then converted into (R)-methylmalonyl-CoA by the action of a racemase enzyme. Then (R)-methylmalonyl-CoA is converted to succinyl-CoA by the action of methylmalonyl-CoA mutase, a vitamin B_{12}-dependent enzyme. Succinyl-CoA is then converted to oxaloacetate and later to malate in the Krebs cycle, also known as citric acid cycle or tricarboxylic acid cycle. Inherited deficiency in propionyl-CoA activity in humans is linked to propionic acidemia with symptoms such as vomiting, lethargy, ketoacidosis, delayed growth, cardiomyopathy, mental retardation, and death in severe cases. Such symptoms may appear in the neonatal period. A large number of autosomal recessive mutations in both subunits of propionyl-CoA carboxylase have been identified, including missense mutations, a few nonsense mutations, insertions/deletions, and splicing mutations.[31] Biotin deficiency causes reduced enzymatic activity of propionyl-CoA carboxylase, which results in the formation of 3-hydroxypropionic acid and 2-methylcitric acid.

The 3-MCC catalyzes conversion of 3-methylcrotonyl-CoA into 3-methylglutaconyl-CoA, a step essential for the catabolism of leucine. Loss of 3-MCC enzyme activity is linked to 3-methylcrotonylglycinuria, an inborn error of metabolism. A large number of autosomal recessive mutations in both subunits of the enzyme have been described.[32] Biotin deficiency may also cause reduced activity of 3-MCC leading to the formation of 3-hydroxyisovaleric acid and 3-methylcrotonyl glycine.[33] Biological activities of biotin-dependent carboxylases are summarized in Table 2.4.

As mentioned earlier, carboxylases are synthesized as apocarboxylases without biotin and the active form is produced by covalent binding of these enzymes to biotin, a reaction catalyzed by holocarboxylase synthetase. After performing their catalytic activities that are essential for life, biotin is released from the ε-amino group of lysine residues of carboxylases in the last step in the degradation of carboxylases by the action of the biotinidase enzyme. In this manner, free biotin is recycled.

Regulation of Chromatin Structure and Genetic Functions by Biotin

The classical role of biotin in metabolism is to serve as a covalently bound coenzyme for five carboxylases. However, there are emerging evidences that biotin participates in processes other than classical carboxylation reactions. Studies have shown novel roles for biotin in cell signaling, gene expression, and chromatin structure. The activity of cell signals such as biotinyl-AMP, Sp1 and Sp3, nuclear factor κB, and receptor tyrosine kinases depends on biotin supply. Consistent with a role for biotin and its catabolites in modulating these cell signals, more than 2000 biotin-dependent genes have been identified in various human tissues. Posttranscriptional events related to ribosomal activity and protein folding may further contribute to the effects of biotin on gene expression.[34]

In eukaryotic nuclei, DNA is grouped into a compact structure to form nucleosomes, the fundamental units of chromatin that also contains the DNA-binding proteins histones and nonhistones. Histones play a very important role in the folding of DNA in the chromatin structure. Five major classes of histones have been characterized in mammals: linker histone H1 and core histones H2A, H2B, H3, and H4. Each nucleosome consists of 146 base pairs of DNA wrapped in an octamer of core histones (one H3-H3-H4-H4 tetramer and two H2A-H2B dimer). Another histone H1 linker is located at the outer surface of each nucleosome and acts as an anchor to fix DNA around the histone core. Chemical modification of DNA and histone (epigenetic changes) affects both the folding of chromatin and its functions. Emerging evidences indicate that distinct lysine residues in both N-terminal and C-terminal regions of human histones are modified by the covalent attachment of biotin, a reaction catalyzed by biotinidase and holocarboxylase synthetase. In fact, holocarboxylase synthetase mediates the binding of biotin to lysine (K) residues in histones H2A, H3, and H4. The existence of biotinylated histone H2A in vivo was confirmed by using modification-specific antibodies. Antibodies to biotinidase and holocarboxylase synthetase localized primarily in the

TABLE 2.4 Biological Activities of Biotin-Dependent Carboxylases			
Enzyme	**Location**	**Reaction**	**Biological Significance**
Acetyl-CoA carboxylase 1	Cytosol	Catalyzes the binding of bicarbonate to acetyl-CoA forming malonyl-CoA.	Acetyl-CoA carboxylase 1 is a key enzyme in fatty acid synthesis in cytosol.
Acetyl-CoA carboxylase 2	Mitochondria	Catalyzes the conversion of acetyl-CoA to malonyl-CoA using bicarbonate and ATP.	Acetyl CoA carboxylase 2 participates in the regulation of fatty acid oxidation in mitochondria, which is mediated by malonyl-CoA, an inhibitor of fatty acid transport to mitochondria.
Pyruvate carboxylase	Mitochondria	Catalyzes the ATP-dependent incorporation of bicarbonate into pyruvate-producing oxaloacetate.	Pyruvate carboxylase is a key enzyme in gluconeogenesis. Oxaloacetate can eventually be converted into glucose.
Propionyl-CoA carboxylase	Mitochondria	Catalyzes the production of methyl malonyl-CoA from propionyl-CoA. However, during biotin deficiency, 3-hydroxypropionic acid and 2-methylcitric acid are produced.	Propionyl-CoA carboxylase catalyzes an essential step in the catabolism of isoleucine, valine, methionine, and threonine; the cholesterol side chain; and odd chain fatty acids.
3-Methylcrotonyl-CoA carboxylase	Mitochondria	Catalyzes the production of 3-methylglutaconyl-CoA from methylcrotonyl-CoA. However, during biotin deficiency, 3-hydroxyisovaleric acid and methylcrotonyl glycine are produced.	Methylcrotonyl-CoA carboxylase catalyzes an essential step in the catabolism of leucine.

TABLE 2.5 Biotinylation Sites in Histones	
Histone	**Biotinylation Sites**
H2A	Lysine (K)-9, K-13, K-125, K127, and K129
H3	Lysine (K)-4, K9, K18, and possibly K23
H4	Lysine (K)-8 and K12

nuclear compartment are consistent with a role for these enzymes in regulating chromatin structure. Biotinylation sites may include N- and C-terminal amino acid residues.[35,36] Biotinylation sites in various histones are listed in Table 2.5.

Biotinylated histones, H1, H2A, H2B, H3, and H4, were initially detected in human lymphocytes in vivo. Later, biotinylated histones have been detected in human lymphoma cells and others.

Biotinylation of histones appears to play a role in cell proliferation, gene silencing, and the cellular response to DNA repair, thus providing genomic stability.[37–39] Interestingly, biotinylation of histones depends on dietary biotin supply.[40,41] Biotinylation of histones is a reversible modification of histones but the mechanism of debiotinylation of histones is not clear. Studies have suggested that biotinidase may catalyze both biotinylation and debiotinylation of histones. Interestingly, the rates of debiotinylation of histones by human plasma and lymphocytes were decreased in biotinidase-deficient specimens.[42]

NUTRITIONAL BIOTIN DEFICIENCY

Biotin deficiency is usually characterized by alopecia and scaly erythematous dermatitis distributed around the body orifices, acidemia, aciduria, hearing and vision problems, and developmental delay in children. Biotin deficiency may also cause paresthesias, myalgias, and mild depression. Biotin deficiency may have adverse effects on the immune system and lipid metabolism. Frank biotin deficiency is rare. However, as discussed earlier, prolonged consumption of raw eggs may cause biotin deficiency because raw egg white contains an antimicrobial protein known as avidin that tightly binds biotin and prevents its absorption. In adults and adolescents who chronically consume raw egg white the following symptoms may develop due to egg white injury syndrome[33]:

- Thinning hair, often with loss of hair color
- Skin rash described as seborrheic and eczematous
- Depression, lethargy, hallucination, and paresthesias of extremities

Pregnancy may cause subclinical biotin deficiency in healthy women because rapidly dividing cells of the developing fetus require biotin for synthesis of essential biotin-dependent carboxylases and also for histone biotinylation. Approximately 50% of pregnant women have an abnormally increased urinary excretion of 3-hydroxyisovaleric acid, which probably reflects decreased activity of the biotin-dependent enzyme 3-methycrotonyl-CoA carboxylase. Mock et al. compared concentrations of biotin and its metabolites as well as 3-hydroxyisovaleric acid in 13 urine specimens obtained from pregnant women (during both early and late pregnancy) with those in urine specimens obtained from 12 nonpregnant women and observed that during early pregnancy, biotin excretion was not significantly different from controls but excretion of 3-hydroxyvaleric acid was increased relative to controls. From early to late pregnancy, biotin excretion decreased in 10 out of 13 women; by late pregnancy, biotin excretion was less than normal in 6 women. During late pregnancy, 3-hydroxisovaleric acid levels remained significantly elevated in the urine of pregnant women compared with controls. Although serum biotin concentrations were significantly greater than those of controls during early pregnancy, serum biotin levels in pregnant women decreased in each woman studied from early to late pregnancy. The authors concluded that biotin status decreases during pregnancy.[43]

In another study based on 26 pregnant women with increased urinary excretion of 3-hydroxyisovaleric acid, the authors observed that supplementation of 300 μg of biotin per day for 14 days reduced urinary excretion of 3-hydroxyisovaelric acid in women both during early and late pregnancy. Therefore a substantial proportion of pregnant women is marginally biotin deficient. Because marginal biotin deficiency during pregnancy does not produce any symptoms of biotin deficiency, the authors commented that widespread biotin supplementation during pregnancy is not currently justified.[44] However, in mice the degrees of biotin deficiency that are metabolically similar to those seen in pregnant women are very teratogenic.[45] Interestingly, in one study based on 504 pregnant women, the authors observed that intake of pantothenic acid, biotin, and magnesium in diet during pregnancy was correlated with increase in infant birthweights.[46]

Prolonged parenteral nutrition without biotin supplementation and infants fed on formula devoid of biotin may cause biotin deficiency. Two adult patients receiving total parenteral nutrition on a long-term home care presented with severe hair loss. Both patients had extensive gut resection, consumed no biotin orally, and received no biotin parenterally. Supplementation with 200 μg per day resulted in gradual regrowth of healthy hair. Then patients were switched to parenteral formula containing biotin and no recurrence of alopecia was observed.[47]

Case Report: A 54–year-old woman with short bowel syndrome was supported with home parenteral nutrition. Six months after receiving 2200 kcal/day of balanced home parenteral nutrition without biotin, she developed biotin deficiency as evidenced by complete hair loss, eczematous dermatitis with waxy pallor, lethargy, and hyperesthesias. Her serum biotin level was 332 pg/mL and urine biotin level was 5.22 ng/mg of creatinine indicating biotin deficiency. The same parenteral nutrition was continued with addition of 10 mg/day oral biotin supplementation. After 3 weeks of intervention, serum (650 pg/mL) and urine (35.6 ng/mg of creatinine) biotin concentrations returned to normal levels. New hair growth was evident and all of her other symptoms resolved. Intravenous biotin was then provided (5 mg/day) for a month after which serum biotin level was increased to 1316 pg/mL and urine biotin level to 178 ng/mg of creatinine. Later the patient had been subsequently maintained on an intravenous multivitamin product containing only 60 μg of biotin per day and the patient remained symptom free indicating no biotin deficiency.[48]

When an infant is weaning from breast and feeding formula, biotin deficiency is rarely observed. However, a 5-month-old Japanese infant who had been diagnosed during neonatal period with dyspepsia, developed typical skin lesions after feeding with amino acid formula that did not contain biotin. Urinary excretion levels of 3-hydroxyisovaleric acid, 3-methylcrotonylglycine, and methylcitric acid were elevated confirming biotin deficiency in the infant. Oral supplement of 1 mg biotin daily corrected biotin deficiency in the infant and organic aciduria was also resolved.[49] Amino acid formula and hydrolyzed formulas given to infants in Japan with milk allergies theoretically contain little if any biotin or carnitine. Hayashi et al. reported cases of six infants with milk allergies who were fed amino acid formula and/or hydrolyzed formula who showed elevated urinary levels of 3-hydroxyisovaleric acid and lower serum concentrations of free carnitine compared with age-matched infants who were fed breast milk or standard infant formulas, indicating biotin and carnitine deficiency in infants fed with amino acid/ hydrolyzed formula. Supplementation with biotin and L-carnitine corrected such deficiency. The authors concluded that care should be taken to avoid biotin and carnitine deficiency in allergic infants fed with amino acid or hydrolyzed formulas.[50] There is a report of biotin deficiency in a girl with type 1B glycogen storage disease caused by an exclusive feeding of a glucose-containing glycogen storage disease formula.[51] A Japanese study involving 46 preterm infants indicates that chronic biotin deficiency may be observed in preterm infants despite feeding maternal milk and/or standard infant formula.[52] Some clinical and biochemical manifestations of biotin deficiency may also occur in severe protein-energy malnutrition. In one study the authors reported that the average plasma biotin concentrations were lower in 16 malnourished children than in 31 controls.[53] Therefore malnourished children from developing countries are at higher risk of biotin deficiency.

Biotinidase is the enzyme responsible for liberating the vitamin biotin from biocytin and dietary protein–bound vitamin. Individuals lacking biotinidase activity become biotin deficient. Because the liver is the major source of plasma biotinidase, chronic liver diseases may decrease serum biotinidase activity and cause biotin deficiency. In one study, using a spectrophotometric method the authors measured serum biotinidase activities in sera from 62 children with chronic liver diseases and from 27 healthy controls. The authors observed normal serum biotinidase activity in patients with noncirrhotic chronic liver diseases (chronic viral hepatitis, prehepatic portal hypertension, glycogen storage disease, Gaucher disease). However, serum biotinidase activities in patients with cirrhosis and Wilson disease were significantly less than those of the control group. The lowest enzyme activities were detected in patients with fulminant hepatitis. The authors concluded that serum biotinidase activity was significantly lower in patients with cirrhosis, particularly in the patients with decompensated cirrhosis and fulminant hepatitis but these patients exhibited no clinical symptoms related to biotin deficiency. However, decreased serum biotinidase activity in chronic liver diseases was associated with severe impairment of hepatocellular function.[54]

Marginal biotin deficiency may occur in women smokers owing to the increased catabolism of biotin by cigarette smoke.[55] Chronic alcohol consumption in humans may cause reduction in plasma biotin levels.[56] This finding has also been reproduced in chronically alcohol fed rats. Inhibition of intestinal biotin absorption by chronic alcohol feeding has been demonstrated in rat model.[57] Moreover, malnutrition and deficiencies of multiple vitamins have been widely reported in alcoholics.[58]

Ketogenic diet (low carbohydrate and high fat) has been reported to cause biotin deficiency in mice. Ketogenic diet increases biotin bioavailability and consumption and hence it promotes energy production by gluconeogenesis and branched-chain amino acid metabolism, which results in biotin deficiency. Therefore biotin supplementation is important for mice after feeding with ketogenic diet.[59] There is a case report of concurrent nonketotic hyperglycinemia and propionic acidemia in an 8-year-old boy who was placed on a ketogenic diet. At 2 years of age, he was diagnosed with nonketotic hyperglycinemia by observing elevated glycine levels and mutations in the GLDC gene. At 8 years of age, after having been placed on ketogenic diet, he became lethargic and had severe metabolic acidosis with ketonuria. Urine organic acid analysis and plasma acylcarnitine profile were consistent with propionic acidemia. He was found to have an apparently homozygous mutation in the PCCB gene: c.49C>A; p.Leu17Met. The patient was treated with natural protein restriction, carnitine, biotin, and thiamine, which resulted in subjective and biochemical improvement.[60] As mentioned earlier, prolonged treatment with anticonvulsants may cause biotin depletion in serum. Patients taking antibiotic for a long time theoretically may experience biotin deficiency because antibiotics may cause alteration of gut flora that produce biotin. Biotin is reabsorbed by the kidney, so studies have addressed potential biotin deficiency in patients suffering from chronic renal failure. However, no biotin deficiency has been reported in such patients. Therefore there is no reason for a regular biotin supplementation in patients with chronic renal failure.[61] However, biotin may be effective in the management of uremic neurologic disorders. In one study the authors supplemented nine patients undergoing chronic hemodialysis for 2–10 years and suffering from encephalopathy (dialysis dementia) and peripheral neuropathy with 10 mg of biotin/day for 1–4 years. Within 3 months, there was a marked improvement in all patients in respect to disorientation, speech disorders, memory failure, myoclonic jerks, flapping tremor, restless legs, paresthesia, and difficulties in walking.[62] The various causes of biotin deficiency are summarized in Table 2.6.

TABLE 2.6
Various Causes of Biotin Deficiency

Cause of Biotin Deficiency	Comments
Egg white injury syndrome	Consuming raw eggs on a regular basis for an extended time causes biotin deficiency because avidin present in egg white tightly binds biotin thus making biotin unavailable for intestinal absorption. Cooking eggs denatures avidin, thus impairing its ability to tightly bind biotin.
Normal pregnancy	Marginal biotin deficiency without any symptoms of frank biotin deficiency may be common during normal pregnancy affecting more than 50% pregnant women. Clinical implication of such marginal deficiency, if any, is not understood. Currently, there is no recommendation for biotin supplementation during normal pregnancy.
Prolonged parenteral nutrition without biotin	Biotin deficiency can be corrected easily by biotin supplementation. Symptoms usually disappear within a few weeks.
Infants fed on formula devoid of biotin	Biotin deficiency can be corrected easily by biotin supplementation. Symptoms usually disappear within a few weeks. However, breastfeeding and feeding with standard infant formulas rarely cause biotin deficiency except in preterm infants. Malnourished children in developing countries are also at risk of developing biotin deficiency.
Severely impaired liver function due to cirrhosis	Reduced biotinidase activity has been reported in patients with severe liver disease but symptoms of biotin deficiency were not observed
Women smokers	Marginal biotin deficiency may occur due to increased catabolism of biotin in smokers.
Treatment with anticonvulsants	Treatment with anticonvulsants such as phenytoin and carbamazepine may cause biotin depletion in serum.
Chronic alcohol abuse	Alcoholics often show deficiencies of various vitamins and minerals, including biotin.
Lipoic acid	Lipoic acid (also known as α-lipoic acid) competes with biotin for binding with sodium-dependent multivitamin transporter, thus potentially decreasing the cellular uptake of biotin.
Ketogenic diet	Mice fed with ketogenic diet develop biotin deficiency.

INBORN ERRORS OF METABOLISM CAUSING BIOTIN DEFICIENCY

Certain inborn errors of metabolism also cause biotin deficiency. The main disorders causing biotin deficiency include

- biotinidase deficiency
- multiple carboxylase deficiency
- isolated carboxylase deficiency
- biotin transport deficiency

Biotinidase Deficiency

Biotinidase deficiency is an autosomal recessively inherited neurocutaneous disorder. Biotinidase deficiency causes biotin deficiencies through multiple mechanisms, including decreased absorption of biotin from the intestine. The lack of biotinidase activity prevents the release of biotin bound to protein in food particles, thus preventing absorption of biotin in the intestine. In addition, poor biotinidase activities also impair recycle of biotin from carboxylases and histones. Typically, symptoms of biotinidase deficiency appear at 1 week after birth to 10 years of age (mean age of onset, 3.5 months). Between January 24, 1984, and December 31, 1988, Wolf and Heard conducted a worldwide survey to establish the incidence of biotinidase deficiency in newborns using screening data of 4,396,834 newborns from 12 countries (Australia, Austria, Canada, Italy, Japan, Mexico, New Zealand, Scotland, Spain, Switzerland, the United States, and West Germany). Biotinidase deficiency was detected in 72 newborns; 32 had profound biotinidase deficiency (less than 10% of mean normal activity level) and 40 had partial deficiency (10%–30% of mean normal activity level). The authors concluded that profound deficiency occurred in 1 per 137,401 live births and partial deficiency in 1 per 109,921 live births, with combined incidence (profound and partial) was 1 case per 61,067 live births. Six children with profound deficiency were symptomatic at, or soon after, the time of diagnosis; but no infant with partial deficiency showed any symptoms of biotin deficiency. Most children whose biotinidase deficiency was detected by newborn screening were Caucasians. However, biotinidase deficiency was also detected in one black child and one Hispanic child. However, no biotinidase deficiency has been detected in Oriental children.[63] There is a case report of a 3-year-old Indian boy presenting with delayed developmental milestones, tachypnea, progressively increasing ataxia, alopecia, and dermatitis due to biotinidase deficiency. However, all his symptoms dramatically responded to high dosages of biotin (30 mg/day).[64]

Biotinidase deficiency can be detected in all affected infants by newborn screening, which is routinely performed in the United States and in many countries. However, in some developing countries, such tests may not be readily available. Children with profound biotinidase deficiency, if untreated, usually exhibit one or more of the following symptoms: hypotonia; seizures; eczematous skin rash; alopecia; respiratory problems, such as hyperventilation, laryngeal stridor, and apnea; conjunctivitis; candidiasis; ataxia; developmental delay; hearing loss; and vision problems, such as optic atrophy. The most common neurologic features of individuals with untreated, profound biotinidase deficiency are seizures and hypotonia. Biotinidase deficiency can be easily treated with 5–10 mg daily oral biotin supplement. The gene for human biotinidase (BTD) is located on chromosome 3q25. The biotinidase gene has been cloned and sequenced; its genomic organization has been determined and more than 150 mutations have been identified that may cause biotinidase deficiency.[65]

Multiple Carboxylase Deficiency

Multiple carboxylase deficiency is a rare autosomal recessive metabolic disease caused by either holocarboxylase synthetase enzyme responsible for covalent binding of biotin with inactive apocarboxylases or biotinidase deficiency. Holocarboxylase synthetase deficiency is usually neonatal or early onset with clinical presentation including severe metabolic acidosis, feeding and breathing difficulties, hypotonia, and lethargy. Some patients may show dermatologic symptoms including erythematous rashes and hair loss. Typical laboratory findings include metabolic acidosis, lactic acidosis, ketosis, hyperammonemia, and organic aciduria showing a typical urinary pattern of excretion of 3-hydroxyisovaleric acid, 3-methylcrotonylglycine, methylcitrate, and 3-hydroxypropionate.[66] Biotin-responsive multiple carboxylase deficiencies are classified into early and late forms. Patients with early form of the deficiency show higher urinary excretion of 3-hydroxyisovalerate and 3-hydroxypropionate than those with the late form. It is proposed that holocarboxylase synthetase is defective in early form of the disease, while impaired intestinal biotin absorption is associated with late form of the disease.[67]

Although 10 mg biotin supplement usually improve clinical symptoms, some patients may only partially respond to biotin therapy and may require much higher doses. Reliable diagnosis of multiple carboxylase deficiency requires enzyme active analysis or preferably genetic analysis. Holocarboxylase synthetase enzyme is encoded by a 11-exon gene, HLCS, located on chromosome 21q22.1. At least 35 mutations of HLCS have been reported.[68] Suzuki et al.[69] suspected that incidence of holocarboxylase synthetase deficiency in Japan is less than 1 in 100,000 live births.

Bailey et al. commented that severe neonatal form of multiple carboxylase deficiency characterized by severe lactic acidosis, cardiovascular compromise, and encephalopathy results from defective holocarboxylase synthetase. Such disorder may be fatal if untreated with biotin supplement. Most HLCS gene mutations causing multiple carboxylase deficiency maps to C-terminal catalytic region of the enzyme and such conditions respond well to therapy with 10 mg per day of biotin. However, patients with some mutations that map to the N-terminal region of holocarboxylase synthetase, outside the proposed catalytic domain, may not respond to biotin therapy.[70]

Case Report: A baby girl weighing 2500 g was born after an uneventful pregnancy and was breast-fed until 5 months of age at which time parents observed an irritability of the baby after ingestion of cow's milk. At the age of 20 months and later at 26 months, she was admitted to the local hospital for lethargy and tachypnea after diarrhea and vomiting. Her metabolic acidosis responded to intravenous fluid and bicarbonate therapy. The patient was admitted again to the hospital at 52 months of age when laboratory findings showed high serum lactate and low blood glucose levels. Her acidosis was corrected by bicarbonate and a vitamin trial with 5 mg of biotin/day and 150 mg/day thiamine was effective but vitamin therapy was stopped at discharge. Finally the diagnosis of holocarboxylase synthetase deficiency was made only at the age of 5.5 years during a metabolic workup when organic acid analysis was performed that revealed elevated levels of urinary excretion of 3-hydroxypropionate, 3-hydroxyisovalerate, and methylcitrate, suggesting multiple carboxylase deficiency that was also demonstrated in the fibroblasts of the patient, but only when the cells were grown in a medium with a very low biotin concentration. The child responded to 10 mg/day of biotin with normal lymphocyte carboxylase activities, and the child showed adequate school performance at 10 years of age.[71]

With oral daily biotin supplementation (10–20 mg), most cases of holocarboxylase synthetase deficiency can be successfully treated, but patients who poorly respond to even very high doses of biotin supplement have a poor long-term prognosis. Of the many discrete gene mutations, errors between amino acids 159 and 314 in the N-terminal extension outside the biotin-binding domain of holocarboxylase synthetase may be associated with disease unresponsiveness to biotin therapy. The L216R mutation seen in individuals of Polynesian ancestry is considered as a lethal mutation because the homozygous mutation is associated with high infant mortality, despite treatment with 100 mg/day of biotin supplement. Wilson et al. describe six patients with homozygous L216R mutation from a family of seven affected children. Four of these children died between the ages of 3 days and 3 years, one had a good recovery after birth but had recurrent infections and metabolic acidosis at 18 months on 40–80 mg of biotin daily, and one was lost to follow-up at the age of 18 months.[72] In contrast to these patients with biotin unresponsiveness, patients heterozygous for a biotin-unresponsive allele and a biotin-responsive allele show good clinical response to biotin therapy between 20 and 40 mg/day. Slavin et al.[73] reported successful therapy with high-dose biotin in surviving siblings with homozygous L216R mutation and commented that the outcome of holocarboxylase synthetase deficiency due to a homozygous L216R mutation, when diagnosed and treated early with high-level neonatal care and biotin, may not be as severe as previously reported.

Isolated Carboxylase Deficiency

Although rare, isolated carboxylase deficiency has also been reported. Beemer et al. reported isolated deficiency of 3-MCC in leucocytes and cultured fibroblasts of two Vietnamese siblings. Both children excreted massive amounts of 3-methylcrotonylglycine and 3-hydroxyisovaleric acid and did not respond to biotin therapy. Fortunately, apart from an attack of vomiting

leading to subcoma in the elder sibling 4 weeks after arrival in the Netherlands, the children were in good health and showed no signs of delayed mental development.[74] Although only a few cases with severe clinical presentation due to isolated 3-MCC deficiency have been reported, profound hypoglycemia is an uncommon but life-threatening complication in this disorder.[75] Murayama et al. reported the case of a 15-year-old girl with a former clinical diagnosis of cerebral palsy, showing isolated deficiency of 3-MCC. Her symptoms included marked growth retardation from birth, profound mental retardation, tonic seizures, quadriplegia with opisthotonic dystonia, gastroesophageal reflux with poor esophageal peristalsis, and recurrent episodes of aspiration pneumonia. Brain magnetic resonance imaging (MRI) revealed marked brain atrophy, involving both the gray and white matter. Although she did not exhibit acute metabolic decompensation or acute encephalopathy, her neurologic symptoms continuously worsened.[76] Bannwart et al. described the fatal outcome of a newborn baby boy with isolated biotin-resistant 3-MCC deficiency (confirmed in cultured fibroblasts). Levels of urinary excretion of 3-hydroxyisovalerate and of 3-methylcrotonylglycine were persistently increased. The patient died from a cardiac and circulatory failure after a prolonged epileptic attack, with bronchial aspiration. The authors commented that nonresponsiveness of the patient to therapy and the fatal outcome indicates the existence of a severe neonatal variant of this otherwise rather benign genetic enzyme deficiency.[77]

Isolated pyruvate carboxylase deficiency has also been described, which may cause lactic acidemia.[78] Propionic acidemia is a rare autosomal recessive organic aciduria due to defective propionyl-CoA carboxylase, a key enzyme of intermediate energy metabolism. Although propionic acidemia is mostly manifested during the neonatal period, when affected, newborns develop severe metabolic acidosis and hyperammonemia. Laemmle et al. reported a case of a late onset of propionic acidemia in a previously healthy teenager of Hispanic origin, who suffered from acute fatigue and breathlessness. Cardiac investigations revealed severe dilated cardiomyopathy, and biochemical investigation (enzymatic and molecular genetic analysis) established the diagnosis of propionic acidemia. The late onset was probably due to residual enzyme activity of approximately 14% of normal enzymatic activity.[79]

Biotin Transporter Deficiency

There is a case report of a 3-year-old boy with biotin deficiency that was not related to biotinidase, or

TABLE 2.7 Hereditary Causes of Biotin Deficiency	
Biotinidase deficiency	Biotinidase deficiency is an autosomal recessively inherited neurocutaneous disorder. Usually symptoms appear at 1 week to less than 1 year of age. Patients with profound deficiency have less than 10% of normal biotinidase activity, whereas patients with partial deficiency have 10%–30% of normal biotinidase activity. Biotin supplementation (5–10 mg per day) is effective in treating these patients.
Multiple carboxylase deficiency	Most severe form is caused by defective holocarboxylase synthetase, while biotinidase deficiency may also cause this disorder. Holocarboxylase synthetase deficiency is a rare autosomal recessive inborn error of metabolism. Although biotin supplementation (10 mg/day) is effective in treating this disorder, some patients may require high dosages of biotin (40–80 mg/day), while patients with *HLCS* gene mutations that map the N-terminal half of the enzyme may not respond to biotin therapy.
Isolated carboxylase deficiency	Although rare, isolated 3-methylcrotonyl-CoA carboxylase and pyruvate carboxylase deficiency have been reported. Propionic acidemia is a rare autosomal recessive organic aciduria due to defective propionyl-CoA carboxylase.
Biotin transport defect	There is a case report of biotin deficiency due to a novel genetic defect in biotin transport. The patient responded to biotin supplement.

holocarboxylase synthetase, deficiency. The child became acutely encephalopathic at age 18 months and showed increased levels of urinary excretion of 3-hydroxyisovaleric acid, 3-methylcrotonylglycine, triglycine, 3-hydroxypropionic acid, propionylglycine, methylcitric acid, lactic acid, and pyruvic acid, indicating multiple carboxylase deficiency. His symptoms rapidly improved after biotin supplementation (10 mg/day). His serum biotinidase activity and biotinidase gene sequence were normal. Activities of biotin-dependent carboxylases in extracts of both cultured fibroblasts and peripheral blood mononuclear cells were normal. These results provided evidence of a novel genetic defect of biotin transport.[80] Hereditary causes of biotin deficiency are summarized in Table 2.7.

BIOTIN THERAPY FOR DISEASES

Several diseases respond to oral biotin therapy. BBGD, also known as thiamine metabolism dysfunction syndrome 2, is a very rare autosomal recessive inherited neurometabolic disorder. BBGD was first described by Ozand et al. in 10 patients, of whom 8 were Saudi Arabian, 1 was Syrian, and 1 was Yemeni. At onset, it appeared as a subacute encephalopathy, with confusion, dysarthria, and dysphagia with occasional supranuclear facial nerve palsy or external ophthalmoplegia, and progressed to severe cogwheel rigidity, dystonia, and quadriparesis. These symptoms disappeared within a few days if biotin was administered but

reappeared within 1 month if biotin was discontinued. Biotinidase and carboxylases activities were within normal range. The etiology may be related to a defect in the transporter of biotin across the blood-brain barrier.[81] In 2005 the disease was mapped to chromosome 2q36.3 and was found to result from a mutation in the SLC19A3 gene that encodes the human thiamine transporter 2 protein. Subsequently, cases have been reported in patients of different ethnicities, including those of Lebanese, Portuguese, German, Indian, and Japanese origins. Neuroradiologic findings include bilateral abnormal signal intensity in the basal ganglia with swelling during acute crises. Alfadhel et al. reported BBGD in 18 children from 13 families, in whom MRI demonstrated abnormal signal intensity with swelling in the basal ganglia during acute crises and atrophy of the basal ganglia and necrosis during follow-up. One-third of the patients showed recurrence of acute crises while on biotin therapy alone, but after the addition of thiamine, crises did not reoccur. All the patients have a homozygous missense mutation in exon 5 of the SLC19A3 gene. The authors concluded that clinicians should suspect BBGD in any child presenting with subacute encephalopathy, abnormal movement, and MRI findings as described earlier. Both biotin and thiamine are essential for disease management because biotin alone could not prevent the recurrence of crises in some patients.[82] Kassem et al.[83] commented that BBGD is a treatable underdiagnosed disease requiring high-dose biotin (5–10 mg/kg) and thiamine (100–300 mg/day) for successful management. Early diagnosis and effective treatment with biotin and thiamine is correlated with positive outcomes.

Biotin in high dosage (100–300 mg/day) is effective in reducing symptoms of multiple sclerosis. It is also effective in treating brittle finger nails and hair loss if a person is suffering from biotin deficiency. Human studies have shown lower biotin levels in patients with type 2 diabetes. Therefore biotin supplementation may be effective in these patients. These topics are discussed in Chapter 3.

ASSESSMENT OF BIOTIN STATUS

Assessment of biotin status is important in determining rare nutritional deficiency of biotin as well as inborn errors of metabolism causing biotin deficiency. In general, serum biotin levels may not decrease in individuals with biotin deficiency and as a result cannot be used for assessment of biotin status. However, decreased levels of urinary excretion of biotin and bisnorbiotin, a biotin metabolite, may be an early and sensitive indicator of biotin deficiency. Increased levels of urinary excretion of 3-hydroxyisovaleric acid (higher than 3.3 mmol/mol of creatinine), a leucine metabolite that is excreted in increased quantities with deficiency of the biotin-dependent enzyme 3-MCC, is also an early and sensitive indicator of biotin deficiency.[24] Eng et al. commented that urinary excretion of biotin effectively identified biotin-supplemented individuals but did not distinguish biotin-deficient and biotin-sufficient individuals. However, urinary excretion of 3-hydroxyisovaleric acid not only identified some biotin-deficient subjects but also produced some false-negative results. Only the abundance of biotinylated 3-MCC (holo-MCC, molecular mass 83 kDa) and biotinylated propionyl-CoA carboxylase (holo-PCC, molecular mass 80 kDa) measured in lymphocytes allowed for distinguishing biotin-deficient and biotin-sufficient individuals.[84]

Measurement of Biotin and Its Metabolites in Serum and Urine

Some microorganisms such as *Lactobacillus plantarum*, *Lactobacillus casei*, *Escherichia coli C162*, *Saccharomyces cerevisiae*, and *Kloeckera brevis* require biotin for growth and can be used for measuring biotin.[37] These microbial methods are the first methods used for measuring biotin but now there are other more specific methods for biotin measurements. The proteins avidin and streptavidin specifically bind biotin and can be utilized for biotin measurement. Avidin-binding assays generally measure the ability of biotin to compete with [³H]biotin or [¹⁴C]biotin for binding to avidin. In one method the authors used biotinylated recombinant aequorin in the development of a heterogeneous bioluminescence binding assay for biotin. The assay was based on a competition between a biotinylated aequorin conjugate and biotin for the binding sites of avidin immobilized on solid particles.[85] Eagle Biosciences (Nashua, NH) has marketed an enzyme-linked immunosorbent assay (ELISA) for biotin based on avidin-biotin interaction. Avidin is bound on the surface of a microtiter plate. Then biotin-containing specimens or standard and a biotin-alkaline phosphatase conjugate are added to wells of the microtiter plate. After 1 hour incubation at room temperature, the wells are washed with diluted washing solution to remove unbound materials. A substrate solution is added and after 30 min incubation a stop solution is added to inhibit formation of the yellow color. The yellow color is measured spectrophotometrically at 405 nm, and concentration of biotin is indirectly proportional to the color intensity.

Streptavidin has greater specificity for biotin than avidin and is the typical protein of choice in bioassays for biotin. A commercially available ELISA test for biotin (Immundiagnostik, Bensheim, Germany) utilizes interaction between streptavidin and biotin. In this method, samples are preincubated with streptavidin-enzyme conjugate and then each incubation mixture is transferred to wells of a microtiter plate coated with biotinylated albumin. After second incubation the substrate of peroxidase (tetramethylbenzidine) is added, and finally the reaction is stopped by using an acidic solution. The intensity of the color is measured at 450 and 630 nm, and biotin concentration is calculated from standard curve.[86]

Biotin can be analyzed using 4-hydroxyazobenzene-2-carboxylic acid. This compound binds with avidin at its biotin-binding sites to produce a characteristic absorption band at 500 nm. The addition of biotin to this complex displaces 4-hydroxyazobenzene-2-carboxylic acid from avidin-binding site and reduces absorption at 500 nm. Therefore intensity of the signal is inversely proportional to biotin concentration.[37]

Chromatographic methods are also available for analysis of biotin, its metabolites, and 3-hydroxyvaleric acid. Mock et al. used high-performance liquid chromatography (HPLC) with a reverse phase C-18 column to separate biotin and its metabolites in urine followed by quantitation using avidin-binding assay. The urinary excretion of 3-hydroxyisovaleric acid was measured using gas chromatography-mass spectrometry (GC/MS).[87] HPLC separation followed by avidin-binding assay can also be used for analysis of biotin and its metabolites (bisnorbiotin and biotin sulfoxide) in serum. Usually biotin accounts for approximately 50% of avidin-binding substances in serum.[88] Thompson et al. described a solid-phase extraction followed by HPLC method using a reverse phase C-18 column for analysis of biotin. The chromatography run time was 8.5 min. After eluting from the column, biotin was subjected to postcolumn reaction to form a conjugate with streptavidin-fluorescein isothiocyanate, which was then detected by a fluorescence detector. This method was tested with infant formula, medical nutritional products, and vitamin premix samples.[89]

Biotin concentration in serum can be accurately measured using ultra-HPLC (UHPLC) combined with tandem MS (UHPLC-MS/MS). Yagi et al. described a simple, rapid, and selective method for determination of plasma biotin using UHPLC-MS/MS. After protein precipitation with methanol, biotin and stable isotope-labeled biotin as an internal standard (biotin-d_4) were extracted and chromatographed on a pentafluorophenyl stationary phase column (2.1×100 mm, 2.7 μm) under isocratic conditions using 10 mm ammonium formate-acetonitrile (93:7, v/v) at a flow rate of 0.6 mL/min. The total chromatographic analysis required 5 min for each injection. The chromatographic column was heated to 50°C. Detection was achieved using positive electrospray ionization MS. The ion transition selected for detection and quantification of biotin was at m/z 245.1→227.0, while for the internal standard the transition was 249.1→231.0. The calibration curve was linear for biotin concentration in the range of 0.05–2 ng/mL. The method required only 300 μL of plasma and was successfully applied to determine plasma biotin concentrations in both healthy individuals and patients undergoing hemodialysis. The authors observed higher serum biotin concentrations (mean, 0.376 ng/mL) in patients undergoing hemodialysis than in healthy individuals (mean, 0.081 ng/mL). The authors also studied stability of biotin in human serum. At room temperature, biotin was stable for 24 h, but after specimen processing, it was stable for 48 h when stored at 4°C. However, if frozen at −15°C the stability of biotin in specimens can be extended to 67 days.[90]

Perry et al. studied how pregnancy and lactation alter biomarkers of biotin metabolism in women consuming a controlled diet. The authors measured biotin in plasma using a commercially available ELISA test for biotin but used LC-MS/MS for analysis of biotin in urine where biotin along with the internal standard (biotin-d_2) were extracted by using solid-phase extraction followed by chromatographic separation using a C-18

reverse phase column. The gradient mobile phase consisted of solution A (0.1% formic acid in water) and solvent B (0.1% formic acid in 1:1 methanol:acetonitrile). The mass spectrometer was operated with electrospray ionization in positive ion mode, monitoring transition m/z 245→227 for biotin and m/z 247→229 for the internal standard. For analysis of the biotin metabolite bisnorbiotin, the authors also used biotin-d_2 as the internal standard. For analysis of bisnorbiotin, transition m/z 217→199 was monitored.[91]

In order to assess biotin status the level of 3-hydroxyisovaleric acid is often measured in urine. Unlike biotin, which is polar and cannot be analyzed using GC/MS, 3-hydroxyisovaleric acid can be derivatized and analyzed by GC/MS. In addition, 3-hydroxyisovaleric acid can also be analyzed using LC-MS/MS. For GC/MS analysis of 3-hydroxyisovaleric acid, Yu et al. prepared trimethylsilyl derivative using N,O-Bis(trimethylsilyl)trifluoroacetamide and used methylmalonic acid-d_3 as the internal standard. For detection of derivatized 3-hydroxyisovaleric acid, m/z 247 [M-15] was selected, and for identification of derivatized internal standard, m/z 250 [M-15] was selected. The authors achieved quantification by using peak areas.[92]

Horvath et al. published a protocol for the analysis of 3-hydroxyisovaleric acid in human urine using UHPLC combined with tandem MS using 3-hydroxyisovaleric acid-d_8 as the internal standard, which was synthesized in their laboratory. For quantitative analysis, specimens were diluted fourfold using deionized water followed by cooling to 5°C in the autosampler during analysis and 1 μL of each was injected into the instrument. The column (reverse phase C-18) was maintained at 55°C during analysis. The mobile phases were 0.01% formic acid and methanol. The initial mobile phase composition was 0% methanol at the time of injection and was held constant for 1 min. Then the percentage of methanol in the mobile phase composition was increased linearly to 100% over 2 min and then again decreased to 0% over 0.2 min. The retention time for both 3-hydroxyisovaleric acid and internal standard ranged between 2.44 and 2.48 min for all samples analyzed. The negative mode electrospray ionization electrospray MS was used for detection and quantitation in selected reaction monitoring mode. The ion transitions of m/z 117.0 to 59.1 and m/z 125.0 to 61.0 were used for 3-hydroxyisovaleric acid and the internal standard, respectively. The collision energy was optimized for both compounds at 20 V.[93]

Increased plasma and urine concentrations of 3-hydroxyisovaleryl carnitine are also used in assessing biotin status. This results from impairment in

TABLE 2.8
Methods for Assessing Biotin Status by Analysis of Biotin and Related Compounds in Serum or Urine

Method	Comments
Microbial methods	*Lactobacillus plantarum, Lactobacillus casei, Escherichia coli C162, Saccharomyces cerevisiae, Kloeckera brevis,* etc. require biotin for growth and can be used for biotin measurement.
Avidin-binding assays	An ELISA test for biotin based on binding of biotin by avidin is commercially available.
Streptavidin-binding assays	Streptavidin binds biotin more specifically than avidin. An ELISA test based on this principle is commercially available.
4-Hydroxyazobenzene-2-carboxylic acid–based assay	This compound binds with avidin and forms a dye that absorbs light at 500 nm. The addition of biotin to this complex displaces 4-hydroxyazobenzene-2-carboxylic acid from the avidin-binding site and absorption at 500 nm is reduced.
Chromatographic methods	Biotin is a polar molecule and cannot be analyzed by gas chromatography/mass spectrometry. Therefore it must be analyzed using high-performance liquid chromatography separation followed by an avidin-binding assay or direct analysis using liquid chromatography combined with tandem mass spectrometry. However, 3-hydroxyisovaleric acid can be measured using gas chromatography/mass spectrometry after derivatization or directly using liquid chromatography combined with tandem mass spectrometry requiring no derivatization.

ELISA, enzyme-linked immunosorbent assay.

leucine catabolism caused by reduced activity of the biotin-dependent enzyme 3-MCC. The concentration of 3-hydroxyisovaleryl carnitine can be accurately measured in serum by using LC-MS/MS, with 3-hydroxyisovaleryl carnitine-d$_3$ as the internal standard. The chromatographic separation could be achieved by using C-18 reverse phase column (heated to 30°C) under isocratic condition using a mobile phase mixture of 60:40 (V/V) of 0.1% trifluoroacetic acid (TFA) in water/0.1% TFA in methanol delivered at a constant flow rate of 500 μL/min. Chromatographic data was acquired on the LC-MS/MS operating in positive ion mode, with m/z 262.2→85 being monitored for 3-hydroxyisovaleryl carnitine and m/z 265.2→85 monitored for the internal standard.[94] A similar approach can also be adopted for estimating urinary concentration of 3-hydroxyisovaleryl carnitine.[95] Methods for assessing biotin status by analysis of biotin and related compounds in serum or urine are summarized in Table 2.8.

CONCLUSIONS

Biotin also known as vitamin B$_7$ or vitamin H is an important vitamin because five carboxylases enzymes essential for important metabolic functions require biotin as a cofactor. Daily requirement of biotin is only 30 μg, which can be easily achieved from eating a balanced meal. As a result, nutritional biotin deficiency is rare, although almost 50% of healthy pregnant women may have marginal biotin deficiency but showing no symptoms. Inborn errors of metabolism involving biotinidase deficiency, multiple carboxylase deficiency, or defects in biotin transport (one case report published) may cause biotin deficiency; however, most congenital causes of biotin deficiency may be treated with oral biotin supplements. Biotin being a polar molecule cannot be analyzed by GC/MS but HPLC followed by an avidin-binding assay or LC-MS/MS methods can be used for measurement of biotin concentration in both serum/plasma and urine. However, 3-hydroxyisovaleric acid, a biomarker of biotin status, can be analyzed by GC/MS after derivatization, but analysis by LC-MS/MS requires no derivatization.

REFERENCES

1. Wolf B, Feldman GL. The biotin dependent carboxylase deficiencies. *Am J Hum Genet.* 1982;34:600–716.
2. McMahon RJ. Biotin in metabolism and molecular biology. *Annu Rev Nutr.* 2002;22:221–239.
3. Bonjour JP. Biotin in human nutrition. *Ann N Y Acad Sci.* 1985;447:97–104.
4. Baugh CM, Malone JH, Butterworth Jr CE. Human biotin deficiency induced by raw egg consumption in a cirrhotic patient. *Am J Clin Nutr.* 1968;21:173–182.
5. Cammalleri L, Bentivegna P, Malaguarnera M. Egg white injury. *Intern Emerg Med.* 2009;4:79–81.

6. Adhisivam B, Mahto D, Mahadevan S. Biotin responsive limb weakness. *Indian Pediatr.* 2007;44:228–230.
7. Staggs CG, Sealey WM, McCabe BJ, Teague AM, et al. Determination of the biotin content of select foods using accurate and sensitive HPLC/avidin binding. *J Food Compost Anal.* 2004;17:767–776.
8. Wolf B, Heard GS, Secor-McVoy JR, Raetz HM. Biotinidase deficiency: the possible role of biotinidase in the processing of dietary protein bound biotin. *J Inherit Met Dis.* 1984;7:121–122.
9. Wang H, Huang W, Fei YJ. Human placental Na+ dependent multivitamin transporter. *J Biol Chem.* 1999;274:14875–14883.
10. Said HM. Cell and molecular aspects of human intestinal biotin absorption. *J Nutr.* 2009;139:158–162.
11. Said HM. Recent advances in Carrier-mediated intestinal absorption of water-soluble vitamins. *Annu Rev Physiol.* 2004;66:419–446.
12. RL D, White BR, Cederberg RA, Griffin JB, et al. Monocarboxylate transporter 1 mediates biotin uptake in human peripheral blood mononuclear cells. *J Nutr.* 2003;133:2703–2706.
13. Zempleni J, Mock DM. Bioavailability of biotin given orally to humans in pharmaceutical doses. *Am J Clin Nutr.* 1999;69:504–508.
14. Chirapu SR, Ortiz A, McCloud E, Dyer D, et al. High specificity in response of the sodium-dependent multivitamin transporter to derivatives of pantothenic acid. *Curr Top Med Chem.* 2013;13:837–842.
15. Mock DM, Malik MI. Distribution of biotin in human plasma: most of the biotin is not bound to protein. *Am J Clin Nutr.* 1982;56:427–432.
16. McCormick DB, Wright LD. The metabolism of biotin and analogs. In: Florkin M, Stotz EH, eds. *Metabolism of Vitamins and Trace Elements.* Amsterdam: Elsevier; 1971:81–110.
17. Said HM. Biotin: biochemical, physiological and clinical aspects. *Subcell Biochem.* 2012;56:1–19.
18. Bitsch R, Salz I, Hotzel D. Studies on bioavailability of oral biotin doses for humans. *Int J Vitamin Nutr Res.* 1989;59:65–71.
19. Peyro Saint Paul L, Debruyne D, Bernard D, Mock DM, Defer GL. Pharmacokinetics and pharmacodynamics of MD1003 (high-dose biotin) in the treatment of progressive multiple sclerosis. *Expert Opin Drug Metab Toxicol.* 2016;12:327–344.
20. Trueb RM. Serum biotin levels in women complaining of hair loss. *Int J Trichol.* 2016;8:73–77.
21. Koutsikos D, Agroyannis B, Tzanatos-Exarchou H. Biotin for diabetic peripheral neuropathy. *Biomed Pharmacother.* 1990;44:511–514.
22. Sedel F, Papeix C, Bellanger A, Touitou V, et al. High doses of biotin in chronic progressive multiple sclerosis: a pilot study. *Mult Scler Relat Disord.* 2015;4:159–169.
23. Debourdeau PM, Djezzar S, Estival JL, Zammit CM, et al. Life-threatening eosinophilic pleuropericardial effusion related to vitamins B5 and H. *Ann Pharmacother.* 2001;35:424–426.
24. Mock DM. Biotin status: which are valid indicators and how do we know? *J Nutr.* 1999;129(2S suppl):498S–503S.
25. Mock DM, Mock N, Nelson RP, Lombard KA. Disturbances in biotin metabolism in children undergoing long-term anticonvulsant therapy. *Pediatr Gastroenterol Nutr.* 1998;26:245–250.
26. Krause KH, Bonjour JP, Berlit P, Kochen W. Biotin status of epileptics. *Ann N Y Acad Sci.* 1985;447:297–313.
27. Krause KH, Kochen W, Berlit P, Bonjour JP. Excretion of organic acids associated with biotin deficiency in chronic anticonvulsant therapy. *Int J Vitam Nutr Res.* 1984;54:217–222.
28. Zempleni J, Trusty TA, Mock DM. Lipoic acid reduces the activities of biotin-dependent carboxylases in rat liver. *J Nutr.* 1997;127:1776–1781.
29. Yates AA, Schlicker SA, Suitor CW. Dietary reference intakes: the new basis for recommendations for calcium, and related nutrients, B vitamin and choline. *J Am Diet Assoc.* 1998;98:699–706.
30. Bull NL, Buss DH. Biotin, pantothenic acid and vitamin E in the British household food supply. *Hum Nutr Appl Nutr.* 1982;36:190–196.
31. J L, Buss DH. Trace nutrients. 5. Minerals and vitamins in the British household food supply. *Br J Nutr.* 1988;60:413–424.
32. Tong L. Structure and function of biotin dependent carboxylases. *Cell Mol Life Sci.* 2013;70:863–891.
33. Zempleni J, Mock DM. Biotin biochemistry and human requirements. *J Nutr Biochem.* 1999;10:128–138.
34. Zempleni J. Uptake, localization, and noncarboxylase roles of biotin. *Annu Rev Nutr.* 2005;25:175–196.
35. YC C, Camporeale G, Kothapalli N, Sarath G, et al. Lysine residues in N-terminal and C-terminal regions of human histone H2A are targets for biotinylation by biotinidase. 17:225–233.
36. B B, Pestinger V, Hassan YI, Borgstahl GE, et al. Holocarboxylase synthetase is a chromatin protein and interacts directly with histone H3 to mediate biotinylation of K9 and K18. *J Nutr Biochem.* 2011;22:470–475.
37. Zempleni J, Wijeratne SS, Hassan YI. Biotin. *Biofactors.* 2009;35:36–46.
38. Kothapalli N, Camporeale G, Kueh A, Chew YC, et al. Biological functions of biotinylated histones. *J Nutr Biochem.* 2005;17:446–448.
39. Hassan YI, Zempleni J. Epigenetic regulation of chromatin structures and gene functions by biotin. *J Nutr.* 2006;136:1761–1765.
40. Gralla M, Camporeale G, Zempleni J. Holocarboxylase synthetase regulates expression of biotin transporters by chromatin remodeling events at the SMVT locus. *J Nutr Biochem.* 2008;19:400–408.
41. Smith EM, Hoi JT, Eissenberg JC, Shoemaker JD, et al. Feeding Drosophila a biotin-deficient diet for multiple generations increases stress resistance and lifespan and alters gene expression and histone biotinylation patterns. *J Nutr.* 2007;137:2006–2012.

42. Ballard TD, Wolff J, Griffin JB, Stanley JS, et al. Biotinidase catalyzes debiotinylation of histones. *Eur J Nutr.* 2002;41:78–84.

43. Mock DM, Stadler DD, Stratton SL, Mock NI. Biotin status assessed longitudinally in pregnant women. *J Nutr.* 1997;127:710–716.

44. Mock DM, Quirk JG, Mock NI. Marginal biotin deficiency during normal pregnancy. *Am J Clin Nutr.* 2002;75:295–299.

45. Mock DM. Marginal biotin deficiency is common in normal human pregnancy and is highly teratogenic in mice. *J Nutr.* 2009;139:154–157.

46. Watson PE, McDonald BW. The association of maternal diet and dietary supplement intake in pregnant New Zealand women with infant birthweight. *Eur J Clin Nutr.* 2010;64:184–193.

47. Innis SM, Allardyce DB. Possible biotin deficiency in adults receiving long term parenteral nutrition. *Am J Clin Nutr.* 1983;37:185–187.

48. Khalidi N, Wesley JR, Thoene JG, Whitehouse Jr WM, et al. Biotin deficiency in a patient with short bowel syndrome during home parenteral nutrition. *PPEN: J Parenter Enteral Nutr.* 1984;8:311–314.

49. Fujimoto W, Inaoki M, Fukul T, Inoue Y, et al. Biotin deficiency in an infant fed with amino acid formula. *J Dermatol.* 2005;32:256–261.

50. Hayashi H, Tokuriki S, Okuno T, Shigematsu Y, et al. Biotin and carnitine deficiency due to hypoallergenic formula nutrition in infants with milk allergy. *Pediatr Int.* 2014;56:286–288.

51. Ihara K, Abe K, Hayakawa K, Makimura M, et al. Biotin deficiency in a glycogen storage disease type 1b girl fed only with glycogen storage disease related formula. *Pediatr Dermatol.* 2011;28:339–341.

52. Tokuriki S, Hayashi H, Okuno T, Yoshioka K, et al. Biotin and carnitine profiles in preterm infants in Japan. *Pediatr Int.* 2013;55:342–345.

53. Velázquez A, Martín-del-Campo C, Báez A, Zamudio S, et al. Biotin deficiency in protein-energy malnutrition. *Eur J Clin Nutr.* 1989;43:169–173.

54. Pabuçcuoğlu A, Aydoğdu S, Baş M. Serum biotinidase activity in children with chronic liver disease and its clinical significance. *J Pediatr Gastroenterol Nutr.* 2002;34:59–62.

55. Sealey WM, Teague AM, Stratton SL, et al. Smoking accelerates biotin catabolism in women. *Am J Clin Nutr.* 2004;80:932–935.

56. Bonjour JP. Vitamins and alcoholism. V. Riboflavin, VI. Niacin, VII. Pantothenic acid, and VIII. Biotin. *Int J Vitam Nutr Res.* 1980;50:425–440.

57. Subramanya SB, Subramanian VS, Kumar JS, Hoiness R, et al. Inhibition of intestinal biotin absorption by chronic alcohol feeding: cellular and molecular mechanisms. *Am J Physiol Gastrointest Liver Physiol.* 2011;300:G494–G501.

58. Fuchs J. Alcoholism, malnutrition, vitamin deficiencies, and the skin. *Clin Dermatol.* 1999;17:457–461.

59. Yuasa M, Matsui T, Ando S, Ishii Y, et al. Consumption of a low-carbohydrate and high-fat diet (the ketogenic diet) exaggerates biotin deficiency in mice. *Nutrition.* 2013;29:1266–1270.

60. Kruszka PS, Kirmse B, Zand DJ, Cusmano-Ozog K, et al. Concurrent non-ketotic hyperglycinemia and propionic acidemia in an eight year old boy. *Mol Genet Metab Rep.* 2014;1:237–240.

61. Jung U, Helbich-Endermann M, Bitsch R, Schneider S, et al. Are patients with chronic renal failure (CRF) deficient in Biotin and is regular Biotin supplementation required? *Z Ernahrungswiss.* 1998;37(4):363–367.

62. Yatzidis H, Koutsicos D, Agroyannis B, Papastephanidis C, et al. Biotin in the management of uremic neurologic disorders. *Nephron.* 1984;36:183–186.

63. Wolf B, Heard GS. Screening for biotinidase deficiency in newborns: worldwide experience. *Pediatrics.* 1990;85: 512–517.

64. Mukhopadhyay D, Das MK, Dhar S, Mukhopadhyay M. Multiple carboxylase deficiency (late onset) due to deficiency of biotinidase. *Indian J Dermatol.* 2014;59: 502–504.

65. Wolf B. Biotinidase deficiency: "if you have to have an inherited metabolic disease, this is the one to have". *Genet Med.* 2012;14:565–575.

66. Morrone A, Malvagia S, Donati MA, Funghini S, et al. Clinical findings and biochemical and molecular analysis of four patients with holocarboxylase synthetase deficiency. *Am J Med Genet.* 2002;111:10–18.

67. Sweetman L. Two forms of biotin-responsive multiple carboxylase deficiency. *J Inherit Metab Dis.* 1981;4:53–54.

68. Tammachote R, Janklat S, Tongkobpetch S, Suphapeetiporn K, et al. Holocarboxylase synthetase deficiency: novel clinical and molecular findings. *Clin Genet.* 2010;78(1):88–93.

69. Suzuki Y, Yang X, Aoki Y, Kure S, et al. Mutations in the holocarboxylase synthetase gene HLCS. *Hum Mutat.* 2005; 26:285–290.

70. Bailey LM, Ivanov RA, Jitrapakdee S, Wilson CJ, et al. Reduced half-life of holocarboxylase synthetase from patients with severe multiple carboxylase deficiency. *Hum Mutat.* 2008;29:E47–E57.

71. Touma E, Suormala T, Baumgartner ER, Gerbaka B, et al. Holocarboxylase synthetase deficiency: report of a case with onset in late infancy. *J Inherit Metab Dis.* 1999;22:115–122.

72. Wilson CJ, Myer M, Darlow BA, Stanley T, et al. Severe holocarboxylase synthetase deficiency with incomplete biotin responsiveness resulting in antenatal insult in Samoan neonates. *J Pediatr.* 2005;147:115–118.

73. Slavin TP, Zaidi SJ, Neal C, Nishikawa B, et al. Clinical presentation and positive outcome of two siblings with holocarboxylase synthetase deficiency caused by a homozygous L216R mutation. *JIMD Rep.* 2014;12:109–114.

74. Beemer FA, Bartlett K, Duran M, Ghneim HK, et al. Isolated biotin-resistant 3-methylcrotonyl-CoA carboxylase deficiency in two sibs. *Eur J Pediatr.* 1982;138:351–354.

75. Oude Luttikhuis HG, Touati G, Rabier D, Williams M, et al. Severe hypoglycemia in isolated 3-methylcrotonyl-CoA carboxylase deficiency; a rare, severe clinical presentation. *J Inherit Metab Dis.* 2005;28:1136–1138.

76. Murayama K, Kimura M, Yamaguchi S, Shinka T, et al. Isolated 3-methylcrotonyl-CoA carboxylase deficiency in a 15-year-old girl. *Brain Dev*. 1997;19:303–305.

77. Bannwart C, Wermuth B, Baumgartner R, Suormala T, et al. Isolated biotin-resistant deficiency of 3-methylcrotonyl-CoA carboxylase presenting as a clinically severe form in a newborn with fatal outcome. *J Inherit Metab Dis*. 1992;15:863–868.

78. Bartlett K, Ghneim HK, Stirk JH, Dale G, et al. Pyruvate carboxylase deficiency. *J Inherit Metab Dis*. 1984;7(suppl 1):74–78.

79. Laemmle A, Balmer C, Doell C, Sass JO, et al. Propionic acidemia in a previously healthy adolescent with acute onset of dilated cardiomyopathy. *Eur J Pediatr*. 2014;173(7):971–974.

80. Mardach R, Zempleni J, Wolf B, Cannon MJ, et al. Biotin dependency due to a defect in biotin transport. *J Clin Invest*. 2002;109:1617–1623.

81. Ozand PT, Gascon GG, Al Essa M, Joshi S, et al. Biotin-responsive basal ganglia disease: a novel entity. *Brain*. 1998;121:1267–1279.

82. Alfadhel M, Almuntashri M, Jadah RH, Bashiri FA, et al. Biotin-responsive basal ganglia disease should be renamed biotin-thiamine-responsive basal ganglia disease: a retrospective review of the clinical, radiological and molecular findings of 18 new cases. *Orphanet J Rare Dis*. 2013;8:83.

83. Kassem H, Wafaie A, Alsuhibani S, Farid T. Biotin-responsive basal ganglia disease: neuroimaging features before and after treatment. *AJNR: Am J Neuroradiol*. 2014;35:1990–1995.

84. Eng WK, Giraud D, Schlegel VL, Wang D, et al. Identification and assessment of markers of biotin status in healthy adults. *Br J Nutr*. 2013;110:321–329.

85. Feltus A, Ramanathan S, Daunert S. Interaction of immobilized avidin with an aequorin-biotin conjugate: an aequorin-linked assay for biotin. *Anal Biochem*. 1997;254:62–68.

86. Wakabayashi K, Kodama H, Ogawa E, Sato Y, et al. Serum biotin in Japanese children: enzyme-linked immunosorbent assay measurement. *Pediatr Int*. 2016;58(9):872–876.

87. Mock DM, Henrich CL, Carnell N, Mock NI. Indicators of marginal biotin deficiency and repletion in humans: validation of 3-hydroxyisovaleric acid excretion and a leucine challenge. *Am J Clin Nutr*. 2002;76:1061–1068.

88. Mock DM, Lankford GL, Mock NI. Biotin accounts for only half of the total avidin-binding substances in human serum. *J Nutr*. 1995;125:941–946.

89. Thompson LB, Schmitz DJ, Pan SJ. Determination of biotin by high-performance liquid chromatography in infant formula, medical nutritional products, and vitamin premixes. *J AOAC Int*. 2006;89:1515–1518.

90. Yagi S, Nishizawa M, Ando I, Oguma S, et al. A simple and rapid ultra-high-performance liquid chromatography-tandem mass spectrometry method to determine plasma biotin in hemodialysis patients. *Biomed Chromatogr*. 2016;30:1285–1290.

91. Perry CA, West AA, Gayle A, Lucas LK, et al. Pregnancy and lactation alter biomarkers of biotin metabolism in women consuming a controlled diet. *J Nutr*. 2014;144:1977–1984.

92. WM Y, Kuhara T, Inoue Y, Matsumoto I, et al. Increased urinary excretion of beta-hydroxyisovaleric acid in ketotic and non-ketotic type II diabetes mellitus. *Clin Chim Acta*. 1990;188:161–168.

93. Horvath TD, Matthews N, Stratton SL, Mock DM, et al. Measurement of 3-hydroxyisovaleric acid in urine from marginally biotin-deficient humans by UPLC-MS/MS. *Anal Bioanal Chem*. 2011;401:2805–2810.

94. Horvath TD, Stratton SL, Bogusiewicz A, Pack L, et al. Quantitative measurement of plasma 3-hydroxyisovaleryl carnitine by LC-MS/MS as a novel biomarker of biotin status in humans. *Anal Chem*. 2010;82:4140–4144.

95. Horvath TD, Stratton SL, Bogusiewicz A, Owen SN, et al. Quantitative measurement of urinary excretion of 3-hydroxyisovaleryl carnitine by LC-MS/MS as an indicator of biotin status in humans. *Anal Chem*. 2010;82:9543–9548.

CHAPTER 3

Biotin: From Supplement to Therapy

INTRODUCTION

Biotin, also known as vitamin B_7 or vitamin H, is an essential vitamin because it acts as a cofactor for five biotin-dependent carboxylases that play critical roles in the intermediate metabolism of gluconeogenesis, fatty acid synthesis, and amino acid catabolism.[1] Pharmacology and pathophysiology of biotin have been discussed in Chapter 2. The daily requirement of biotin is only 30 µg for both men and women, while daily biotin requirement is much less in infants and children (Table 3.1). Many foods including organ meat such as liver, meat, eggs, fish, seeds, nuts, and certain vegetables such as sweet potatoes contain biotin. Although there is no nationally representative estimate of biotin intake in the United States, the average biotin intake from food in Western population has been estimated to be 35–70 µg/day.[2]

Like other members of the vitamin B complex, biotin is water soluble and not toxic. Multivitamin formulations usually contain approximately 30 µg biotin. Nutritional biotin deficiency is rare but healthy pregnant women may experience marginal biotin deficiency without showing any symptoms. As a result, biotin supplement in pregnant women is not recommended as a routine therapy. However, there is a recent trend to take megadoses of biotin for healthy hair growth, skin, and nails. People often take 5–10 mg of biotin supplement per day for such purposes. Biotin in higher doses may be needed to treat congenital biotin deficiency. Biotin in 100–300 mg dosage per day is used for treating symptoms of multiple sclerosis.

Although normal intake of biotin from diet, taking multivitamin supplements containing biotin, or intake of biotin supplement containing less than 5 mg of biotin per day have no impact on biotin based immunoassays, taking biotin supplement at higher dosages (5–10 mg per day) or pharmaceutical dosage of biotin (100–300 mg/day) may impact clinical laboratory test results using immunoassays that utilize biotinylated antibodies. Usually high concentration of biotin falsely increases test results using competitive immunoassays but falsely lower test results using sandwich immunoassay format. Wrong diagnosis of hyperthyroidism due to false elevated levels of free thyroxine and free triiodothyronine but false lower levels of thyroid-stimulating hormone has been documented in the literature. FDA reported death of a person due to falsely lower troponin values as a result of biotin intake that caused missed diagnosis of myocardial infarction. Please see Chapter 4 for an in-depth discussion on the effect of biotin in clinical laboratory test results using biotin-based immunoassays.

Biotin supplements particularly at large dosages have become increasingly popular, as approximately 15%–20% people in the United States consume biotin-containing supplements.[3] As mentioned in Chapter 4, retail sales of biotin supplements grew more than 260% between 2013 and 2016. The total sales are probably much higher because the Nelson Food Drug sales data do not include sales from wholesale warehouses. This chapter focuses on why people take high-dose biotin supplement and if there is any scientific data to support the claim that biotin supplement is helpful for healthy hair, nails, and skin. In addition, multiple publications indicate beneficial roles of biotin in improving quality of life in patients suffering from multiple sclerosis. Moreover, biotin may also be helpful for maintaining healthy glucose levels in type 2 diabetic patients, as well as improving lipid profile and reducing weight. Biotin may also have favorable effects on immune function and may reduce hypertension. All these potential benefits of biotin are discussed in this chapter.

USE OF BIOTIN SUPPLEMENTS

Major appeal of biotin supplements to the general public is for healthy hair, nails, and skin. Owing to the relatively low cost, biotin supplement has become a new trend for consumers wishing to have longer healthier hair and nails. Moreover, biotin is a water-soluble vitamin with no known toxicity. Usually multivitamin formulations contain 30–60 µg of biotin, which is sufficient to satisfy daily biotin requirements. However, pure biotin supplements in doses of 300 µg to 100 mg are commercially available. Searching amazon.com and looking at various health food stores, the authors found various dosages of biotin that are listed in Table 3.2 with popular intended use as well as potential effects

TABLE 3.1
Biotin Requirements in Humans

Age	Daily Biotin Requirement in Both Men and Women (μg)
Birth to 6 months	5
7–12 months	6
1–3 years	8
4–8 years	12
9–13 years	20
14–18 years	25
19+ years (adults)	30[a]

[a]Although daily requirement of biotin is 30 μg in pregnant women, during lactation the daily requirement is 35 μg.
Based on Biotin fact sheets for health professionals. National Institute of Health: Office of Dietary Supplement.

on laboratory tests. Pharmaceutical dose of 300 mg of biotin is also available by prescription. It is important to know that taking high-dose biotin (more than 5 mg per day) may affect some clinical laboratory test results using biotin-based immunoassays. Please see Chapter 4 for more details.

Biotin Supplement for Healthy Hair, Nails, and Skin

Although current daily requirement of biotin is only 30 μg, which can be easily satisfied by eating a balanced meal, many individuals still take 0.5–1 mg daily biotin supplement. Biotin's function in protein synthesis and more specifically in keratin production explains its role in healthy hair and nail growth. Moreover, signs of biotin deficiency include hair loss, brittle nails, and skin rashes. Although biotin supplementation showed improvement of hair and nail growth in patients with documented biotin deficiency, biotin has no proven

TABLE 3.2
Commercially Available Doses of Biotin Supplements and Their Effect on Biotin-Based Immunoassays

Biotin Dose	Intended Use	Comments
Vitamin B complex and multivitamins: contain 30–60 μg of biotin	Good health	Daily biotin requirement in adults is 30 μg and eating a balanced meal can satisfy the requirement. However, taking multivitamins containing 30–60 μg biotin has no effect on clinical laboratory test results in biotin-based immunoassays.
Biotin with chromium	Glucose control	One study with Diachrome (2 mg biotin and 600 μg chromium) indicates the glucose-lowering effect of this formula. Some supplements for glucose control have biotin, chromium, and cinnamon. Cinnamon may also have glucose-lowering effect.
250 μg	Good health	No effect on clinical laboratory tests.
300 μg	Good health	No effect on clinical laboratory tests.
1000 μg (1 mg)	Good health	No effect on clinical laboratory tests.
2500 μg (2.5 mg)	For hair and nail growth	Scientific evidence indicates that this dose is effective in treating brittle nails. No effect on clinical laboratory tests.
5000 μg (5 mg)	Popular dose recommended for healthy hair, nails, and skin	Currently there is no scientific evidence that biotin helps with hair and nail growth in healthy subjects with no biotin deficiency. Taking biotin at a dosage up to 5 mg/day may not affect clinical laboratory tests.
10,000 μg (10 mg)	Recommended for healthy hair, nails, and skin.	No scientific research supports such claims. However, one study showed reduced fasting sugar levels in diabetic patients who took 9 mg biotin daily. However, taking biotin at 10 mg/day may affect some clinical laboratory test.
15,000 μg (15 mg)	Recommended for healthy hair, nails, and skin.	No scientific research supports such claims. However, taking biotin at 15 mg/day affects some clinical laboratory tests.

TABLE 3.2 Commercially Available Doses of Biotin Supplements and Their Effect on Biotin-Based Immunoassays—cont'd		
Biotin Dose	**Intended Use**	**Comments**
100,000 µg (100 mg)	High-potency biotin	There is no scientific evidence supporting health benefits of 100 mg/day biotin in healthy individuals. However, people with inherited biotin deficiency require high doses of biotin as recommended by the physician. Taking such high doses of biotin affects all biotin-based immunoassays.
100–300 mg Pharmaceutical grade biotin	Under clinical investigation	Preliminary data from clinical trials indicate that high-dose pharmaceutical grade biotin is effective in reducing symptoms of a subset of patients with multiple sclerosis. However, taking such high-dose biotin affects all biotin-based immunoassays.

effect on hair and nail growth in healthy individuals. However, in cases of acquired and inherited causes of biotin deficiency and diseases, such as brittle nail syndrome or uncombable hair syndrome, biotin supplementation may be beneficial.[4]

Limat et al.[5] reported that proliferation and differentiation of cultured human follicular keratinocytes are not influenced by biotin. Therefore biotin has no role in hair and nail growth in healthy individuals without any biotin deficiency. Soleymani et al. commented that although there exists an incredible amount of social media hype and market advertising touting the efficacy of biotin supplement for the improvement of hair quantity and quality, biotin's efficacy for hair remains largely unsubstantiated in scientific literature, and to date, there have been no clinical trials conducted to investigate the efficacy of biotin supplementation for the treatment of alopecia of any kind or any randomized controlled trial to study its effect on hair quality and quantity in human subjects. Therefore societal infatuation with biotin supplementation is propagated not only by its glamorization in popular media but also because its popularity is vastly disproportionate to the insufficient clinical evidence supporting its efficacy in hair improvement.[6]

Although biotin supplement may not improve quality of hair in healthy individuals, people who experience hair loss (alopecia) due to biotin deficiency may receive benefits from oral biotin supplementation. In one study, the authors measured serum biotin levels in 503 women complaining of hair loss and observed that 189 women showed biotin deficiency (serum biotin level <0.1 ng/mL where adequate level is considered as 0.4 ng/mL). However, approximately 11% of patients with biotin deficiency had a positive personal history for risk factors for biotin deficiency. The authors

commented that the custom of treating women complaining of hair loss in an indiscriminate manner with oral biotin supplementation should be rejected, unless biotin deficiency has been established and other causes of hair loss have been ruled out.[7] Patel et al. reviewed 18 cases where all patients received biotin supplementation due to some underlying pathology for either poor hair or nail growth. Out of 18 patients, 10 cases had inherited deficiency of either biotinidase or holocarboxylase synthetase. Of these 10 cases, eight cases reported reversal of hair loss due to biotin supplementation. In addition, three cases of uncombable hair syndrome also responded to oral biotin supplement. One case had secondary biotin deficiency due to diet. In addition, biotin supplement of 2.5–3 mg/day resolved brittle nail syndrome in three cases.[4]

Abnormalities of the hair shaft are associated with unruly hair including uncombable hair syndrome and woolly hair. Uncombable hair syndrome is an inherited autosomal dominant trait or a sporadic structural hair abnormality usually seen in people with blond to light brown hair with a tangled appearance. Woolly hair refers to tightly coiled hair covering the whole scalp or part of it, in a non-negroid individual. These hairs tend to be unruly, curly, lighter in color, and smaller in diameter than the surrounding normal hair without an increase in fragility and appear to be sparse. A 10-year-old girl presented to the authors with a complaint of abnormal hair from the age of 2 years. The child was born with normal black hair all over the scalp. At the age of 2 years, the entire scalp was shaved bald as part of a religious ceremony. The mother noticed that the new hair that grew all over the scalp was light-colored and dry, sticking out of the scalp, and uncombable. However, the child was developmentally normal and there was no personal or family history of cutaneous disease.

On clinical examination, hair on the entire scalp was normal but for a patch on the occipital region that was dry, lighter in color (blond to light brown), sticking out of the scalp, and could not be combed flat. Electrocardiography, echocardiography, routine blood tests, and urine examination were normal. Light microscopy of the uncombable hair showed light brown pigment. Oral biotin, 5 mg once daily, was started. The patient and her parents were counseled about the benign and chronic nature of her condition.[8]

Biotin supplement is effective in treating uncombable hair syndrome, consisting of slow-growing, straw-colored scalp hair that could not be combed flat. The hairs appear normal on light microscopy, but on scanning electron microscopy, they are triangular in cross section, with canal-like longitudinal depressions. In one study the authors treated three children with uncombable hair syndrome with oral biotin, 0.3 mg three times a day, and observed significant improvement in one patient after 4 months. The other two patients had associated ectodermal dysplasia but their hair slowly improved in appearance over 5 years even without biotin therapy.[9] Boccaletti et al. reported a family affected to the fourth generation by uncombable hair syndrome. The family history strongly supported the hypothesis of autosomal dominant inheritance and diagnosis was confirmed by extensive scanning electron microscopy. Therapy with oral biotin 5 mg/day was started on two young patients, with excellent results as per as appearance of hair was considered. After a 2-year-period of follow-up, hair normality was maintained without biotin, while nail fragility still required biotin supplementation.[10] Valproic acid–induced biotin deficiency and subsequent hair loss may also be successfully treated with oral biotin supplementation.

Case Report: MA had bipolar I disorder and had been on lithium monotherapy since age 23 years. After 18 years (with two drug-free intervals followed by manic episodes), at age 42 years, she developed tubular interstitial nephritis when lithium therapy was replaced by carbamazepine monotherapy. However, after 4 years, carbamazepine was substituted by valproic acid due to deterioration of liver functions. Then she developed telogen effluvium 1 month after the introduction of valproate treatment, with no other significant side effects. At 9 months after starting valproate, severe hair loss was observed and she was treated with 10 mg oral biotin supplement daily with good results. Three months later, her excessive hair loss completely disappeared, and she was content to have her usual hair volume again. Twelve months later, the biotin treatment was tapered off at her request. Now, 7 years later, with the patient still successfully on valproate treatment, no recurrence of hair loss has developed. The authors commented that adding biotin can be a simple, safe, and effective treatment option for a subgroup of patients with valproate-induced hair loss.[11]

Mercke et al. commented that alopecia is most commonly induced by some antipsychotic and antianxiety agents although hair loss is also related to hypothyroidism induced by lithium and other agents. Lithium causes hair loss in 12%–19% of long-term users, whereas valproic acid and/or divalproex may cause alopecia in up to 12% of patients in a dose-dependent relationship. Incidences up to 28% are observed with high serum valproate concentrations. These drugs may also change hair color and structure. The occurrence of carbamazepine-induced alopecia is at or below 6%. Hair loss is less common with other psychoactive drugs, including tricyclic antidepressants, maprotiline, and trazodone, and rare with all the new generation of antidepressants. Discontinuation of the medication or dose reduction almost always leads to complete hair regrowth.[12] Ramakrishnappa and Belhekar reported a case of 26-year-old woman who experienced significant hair loss when her serum valproic acid concentration was increased from 70.14 to 124.82 μg/mL (therapeutic range, 50–100 μg/mL). However, her hair loss was reversed by substituting valproic acid with levetiracetam.[13]

Brittle fingernails (onychoschizia) including splitting, frail, soft, or thin nails occur more often in women than men. When questioned, approximately 20% women complained about brittle fingernails. Colombo et al. studied the effect of 2.5 mg/day oral biotin supplement in treating brittle fingernails. Biotin therapy was continued for 6 months and then stopped if improvement was observed. If no improvement was observed, then therapy may be continued for an additional 9 months. The authors investigated the distal ends of the fingernails from 32 persons placed into three groups: group A consisted of 10 control subjects with normal nails; group B, eight patients with brittle nails studied before and after biotin treatment using scanning electron microscopy; and group C, 14 patients with brittle nails in whom the administration of biotin did not coincide exactly with the initial and terminal clipping of the nails. The thickness of the nails in group B increased significantly by 25% but in group C, the increase was 7%. Splitting of the nails was reduced in groups B and C, and the irregular cellular arrangement of the dorsal surface of brittle nails became more regular in the nails of all patients in group B and in those of 8 of 11 patients in group C. The authors concluded that oral biotin supplement of 2.5 mg per day is effective in treating brittle nails.[14] In another study involving 35 patients with brittle nails, the authors observed that oral biotin supplementation of 2.5 mg per day for 6–15 months resulted in clinical improvement in 22 patients (63% response rate).[15] Floersheim[16] observed significant clinical improvement in 41 out of 45 patients with brittle fingernails following 2.5 mg daily oral biotin supplement for an average of 5.5 months of treatment.

Triangular worn down nails are a subset of brittle nails that are characterized by a triangular area of nail plate thinning extending from the mid-nail plate to the distal margin. Usually two to four fingernails in the dominant hand are affected. Piraccini et al. have treated 14 healthy patients with triangular worn down nails affecting several fingernails. Six of them worked as tailors and rubbed their nails against clothes while sewing. All patients were instructed to cut their nails short, and also to avoid the rubbing habit. These patients were treated with a 10% urea–containing ointment and 5 mg/day of oral biotin supplementation. Evaluation after 6 months revealed improvement of the nail abnormalities in nine patients and complete cure in five patients. The follow-up revealed complete cure in all patients. However, recurrences were observed in three patients.[17]

Trachyonychia is the roughness of the nail plate which can be either shiny or opaque type. Oral biotin therapy (2.5 mg/day) was associated with a beneficial effect on trachyonychia in two children with both the opaque and shiny types.[18] Habit-tic deformity is a form of nail dystrophy caused by habitual external trauma to the nail matrix and may affect any nail, but thumbnails are the most commonly affected. This deformity is due to manipulating the proximal nail fold or periungual area with an adjacent fingernail, often in an unconscious manner. Multivitamin supplement containing 6000 µg biotin may help in treating this disorder, but oral biotin (2.5 mg/day) alone was not effective in treating the relapse. However, reintroduction of multivitamin supplement corrected the deformity.[19]

There is limited data on biotin supplementation to treat dermatologic disorders, especially in patients with a normal biotin level. In biotin-deficient children with atopic dermatitis, treatment with oral biotin supplement is effective. Surprisingly, despite inconclusive evidence to support therapeutic benefits of biotin for dermatologic conditions, 66% dermatologists (198 out of 300) surveyed recommended dietary supplement to their patients most commonly to improve their skin, hair, and nails. There is an upward trend of consumers taking biotin supplements as well as physicians recommending biotin supplements, but the FDA warning that taking biotin supplements may significantly interfere with laboratory test results should give everyone a pause.[20]

Biotin Therapy in Multiple Sclerosis

Multiple sclerosis is an autoimmune disease characterized by both progressive damages to the myelin sheath surrounding nerve fibers and neuronal loss in the brain and spinal cord of patients suffering from this disease. Multiple sclerosis affects an estimated 2.3 million people worldwide and is the most common disabling neurologic disease of young adults, with first symptoms typically manifesting between 20 and 40 years of age. The prevalence is more common in women than men. In the majority of patients (approximately 85%) an initial phase of relapsing-remitting neurologic dysfunction typically results in a secondary progressive phase with gradual worsening of neurologic disability leading to problems with vision, walking, balance, incontinence, cognitive changes, fatigue, and pain. Primary progressive multiple sclerosis, characterized by disease progression from onset, is less common, affecting 10%–15% of patients.

Regardless of the pattern of the disease, treatment options for multiple sclerosis remain inadequate. High-dose biotin has emerged as an attractive therapy option to reduce symptoms in patients suffering from multiple sclerosis. MD1003 (MedDay Pharmaceuticals, Paris, France) is an oral formulation of high-dose pharmaceutical grade biotin currently in clinical development for treating progressive multiple sclerosis and adrenomyeloneuropathy. The 300-mg daily dose of biotin is 10,000-fold higher than the recommended daily intake of biotin. Studies have shown that this high-dose biotin therapy is effective in reducing symptoms of multiple sclerosis. Sedel et al.[21] hypothesized that high-dose biotin exerts its therapeutic effect in progressive multiple sclerosis through two primary mechanisms:

- Promoting remyelination through enhanced myelin formation in oligodendrocytes.
- Enhancing brain energy production, thus protecting demyelinated axons from degradation.

In one clinical trial the authors treated 23 consecutive patients with primary and secondary progressive multiple sclerosis by administering high doses of biotin for 2–36 months (mean = 9.2 months). Most patients (n = 19) received 300 mg of biotin per day, divided into three doses, whereas one patient received 600 mg of biotin and another two patients received 200 mg of biotin per day. In four patients with prominent visual impairment related to optic nerve injury, visual acuity improved significantly following high-dose biotin therapy. One patient with left homonymous hemianopia also gradually improved from 2 to 16 months following biotin therapy. In addition, 16 out of 18 patients (89%) with prominent spinal cord involvement also showed improvement after biotin therapy for 2–8 months. Overall, 21 out of 23 patients (91.3%) showed some improvements following high-dose biotin therapy. The dosage of 300 mg of biotin per day was thought to be associated with best clinical results. The treatment appeared to be safe, with transient diarrhea being the only minor adverse effect. The authors concluded from their preliminary results that high-dose biotin therapy may be effective in treating symptoms of multiple sclerosis.[22]

The preliminary results of a multicenter placebo-controlled trial involving 154 patients with multiple sclerosis (103 patients received 300 mg of biotin orally daily and 51 patients received placebo) showed that 13 patients (12.6%) treated with high-dose oral biotin had a reduction in multiple sclerosis–related disability at month 9 and confirmed at month 12, compared with none of the 51 patients receiving placebo. High-dose biotin was well tolerated in all patients. The authors concluded that high-dose biotin therapy was effective in reducing symptoms related to multiple sclerosis in a subset of patients.[23] However, high-dose biotin therapy is associated with significant interference with biotin-based immunoassays (positive interference with competitive format but negative interference with sandwich format), which may cause wrong clinical diagnosis. Wrong diagnosis of hyperthyroidism in euthyroid patients because of interference of high-dose biotin in thyroid function tests using biotin-based immunoassays has been well documented in the medical literature.[24] Please see Chapter 4 for more details.

Biotin Therapy in Diabetic Patients
Experiments with rats indicate that biotin therapy may effectively lower serum glucose levels. Zhang et al. reported that a high-biotin diet improved the impaired glucose tolerance of long-term spontaneously hyperglycemic rats with non–insulin-dependent diabetes mellitus.[25] Biotin probably lowers glucose both by stimulating glucose-induced insulin secretion by the pancreatic beta cells and by increasing glycolysis in the liver and pancreas. Biotin may also improve utilization of glucose by muscles by increasing the guanylate cyclase activity.[26] Maebashi et al. reported lower serum biotin levels in 43 patients (25 males and 18 females; mean age, 46 years; range, 35–56 years) suffering from

type 2 diabetes (fasting glucose > 171 mg/dL, duration of disease, 2–6 years) and who were resistant to treatment with sulfonylureas compared with 64 controls. The mean biotin level in diabetic patients was 56.7 nmol/L, but the mean biotin level in the controls was 96.8 nmol/L. An oral biotin supplement of 9 mg daily (given in three divided doses) was administered to 18 patients but another 10 patients received placebo. In patients who received biotin the mean serum glucose level was reduced from 12.9 mmol/L (232.2 mg/dL) to 7.1 mmol/L (127.8 mg/dL) after 1 month of administration of biotin. However, after 1 month of cessation of biotin therapy, fasting glucose level returned to the initial values. In addition, serum pyruvate and lactate levels were also reduced after biotin therapy but returned to the normal values after cessation of therapy. However, serum insulin levels were not changed during biotin administration. As expected, no reduction in glucose, pyruvate, and lactate levels was observed in patients taking placebo. The long-term administration of biotin was tested in 20 patients, and 5 patients were followed up for more than 4 years. The fasting blood glucose level decreased to normal level within 2 months and remained within the normal range thereafter. However, serum insulin level was almost unchanged.[27]

Available evidence suggests that a number of natural supplements, including cinnamon, biotin, fenugreek, ginseng, and α-lipoic acid, have the potential to reduce the risk for type 2 diabetes and to improve optimal glycemic control in patients with type 2 diabetes.[28] Coggeshall et al.[29] reported 50% reduction in fasting serum glucose levels in diabetic patients after receiving 16 mg biotin daily for 1 week. Biotin in high doses (10 mg per day intramuscularly for 6 weeks followed by 5 mg biotin daily) was given for 1–2 years to three diabetic patients suffering from severe diabetic peripheral neuropathy. Within 4–8 weeks, there was a marked improvement in clinical and laboratory findings indicating that diabetic patients may suffer from biotin deficiency due to inactivity or unavailability of biotin. Reduced activity of the biotin-dependent enzyme, pyruvate carboxylase, may lead to accumulation of pyruvate and/or depletion of aspartate, both of which play a significant role in the nervous system metabolism. The authors suggested regular biotin administration in diabetic patients.[30]

Preclinical studies have shown that the combination of chromium picolinate and biotin significantly enhances glucose uptake in skeletal muscle cells and increases glucose disposal. Singer and Geohas studied effect of 600 μg of chromium as chromium picolinate and biotin (2 mg/day) (Diachrome, Nutrition 21, Inc) supplementation on glycemic control in 43 subjects with impaired glycemic control (2-h glucose >200 mg/dL, glycated hemoglobin≥7%), despite treatments with oral antihyperglycemic agents. The patients were randomized to either receive the supplement or placebo. The authors reported that after 4 weeks, there was a significantly greater reduction in the total area under the curve for glucose during the 2-h oral glucose tolerance test for the treatment group than the placebo group. Significantly greater reductions were also seen in fructosamine, triglycerides, and triglycerides/high-density lipoprotein cholesterol ratios in the treatment group. The authors reported no adverse effects during therapy with chromium picolinate and biotin and concluded that chromium picolinate/biotin supplementation may represent an effective adjunctive nutritional therapy to people with poorly controlled diabetes.[31] However, in one study involving 10 patients with type 2 diabetes and 7 control subjects, the authors observed no significant change in glucose, insulin, triacylglycerol, cholesterol, or lactate concentration following treatment with biotin supplement (5 mg three times a day) in either the diabetic or the control subjects.[32] However, in a study using mice model the authors observed that feeding mice for 8 weeks with biotin-rich diet resulted in increased insulin secretion together with the messenger RNA (mRNA) expression of several transcription factors regulating insulin expression and secretion, including forkhead box protein A2, pancreatic and duodenal homeobox 1, and hepatocyte nuclear factor 4α. The mRNA abundance of glucokinase, CACNA1D, acetyl-CoA carboxylase (ACC), and insulin also increased in biotin-fed mice. The authors concluded that these findings provide, for the first time, an insight into how biotin supplementation exerts its effects on the function and proportion of beta cells, suggesting a role for biotin in the prevention and treatment of diabetes.[33]

Biotin for Lowering Lipid Levels

Biotin reduces triglyceride concentrations in both human and animal models. In one study using a mice model the authors investigated signaling pathways and posttranscriptional mechanisms involved in the hypotriglyceridemic effects of biotin. The authors observed that compared with the control group, biotin-supplemented mice had lower serum and hepatic triglyceride concentrations. Biotin supplementation increased the levels of cyclic guanosine monophosphate (cGMP) as well as the phosphorylated forms of 5′-adenosine monophosphate–activated protein kinase (AMPK) and ACC-1 while decreasing the abundance of the mature form of sterol regulatory element–binding protein 1c (SREBP-1c; regarded as a major factor

involved in the nutritional regulation of lipogenesis) and fatty acid synthase (FAS). Therefore biotin reduces lipogenesis by increasing cGMP content and AMPK activation, which results in augmented ACC-1 phosphorylation and decreased expression of both the mature forms of SREBP-1c and FAS.[34]

Larrieta et al. also studied the molecular mechanism of the lipid lowering effect of biotin using mice model. Three-week-old male BALB/cAnN Hsd mice were fed a biotin-control or a biotin-supplemented diet (1.76 or 97.7 mg of free biotin/kg diet, respectively) over a period of 8 weeks. The authors observed 35% reduction in serum triglyceride levels in biotin-fed mice compared with the control group. Moreover, in the liver the authors observed a significant reduction of mRNA levels of SREBP-1c, glucose transporter 2, phosphofructokinase 1, pyruvate kinase, ACC, and FAS, while glucose-6-phosphate dehydrogenase expression increased. In adipose tissue, decreased expression of SREBP-1c, glucose-6-phosphate dehydrogenase, ACC, FAS, stearoyl-CoA desaturase 1, phosphofructokinase 1, and peroxisome proliferator–activated receptor γ was also observed. Interestingly, adipose tissue weights were significantly decreased in mice fed with biotin compared with the control group. The authors concluded that pharmacologic concentrations of biotin decrease serum triglyceride concentrations and lipogenic gene expression in liver and adipose tissues.[35]

Favorable effect of biotin supplement in improving lipid profile has also been reported in studies involving human subjects. Marshall et al. conducted a double-blind study with 40 men and women volunteers aged 30–60 years to investigate the effect of biotin supplementation (0.9 mg/day) for 71 days on plasma lipids and other plasma constituents. The authors observed the largest differences during the first 2 weeks after supplementation when small but statistically significant positive changes for lipid profile were observed in biotin-treated men and women compared with the control population. At the end of the study, these levels were at or below initial levels. Moreover, there was a negative correlation between biotin levels and total plasma lipids. Interestingly, subjects with elevated lipid levels at the beginning of the study showed higher reductions in lipid profile than subjects with normal lipid profile during initiation of the study.[36] Therefore biotin supplement may be effective in improving lipid profile.

In another study the authors investigated the effect of biotin administration on the concentration of plasma lipids, as well as glucose and insulin, in type 2 diabetic and nondiabetic subjects. A total of 18 diabetic and 15 nondiabetic subjects aged 30–65 years were randomized into two groups and received 15 mg biotin supplement each day or placebo for 28 days. The authors observed that biotin supplement significantly reduced plasma triglyceride and very-low-density lipoprotein concentrations in both diabetic and nondiabetic subjects but had no significant effects on cholesterol, glucose, and insulin concentrations in either the diabetic or the nondiabetic subjects. The authors concluded that biotin therapy decreases hypertriglyceridemia.[37] A Russian study reported that daily supplementation of 5 mg biotin for 4 weeks reduced hypercholesterolemia in human subjects.[38]

Hemmati et al. conducted a randomized double-blind placebo-controlled clinical trial involving 70 type 1 diabetic patients in the age range 5–25 years and with poorly controlled (glycosylated hemoglobin ≥8%) glycemic control. Subjects were randomly allocated into two groups. In the intervention group (35 patients), biotin (40 µg/kg, but not exceeding 2 mg biotin per day) was administered along with daily insulin, whereas the control group (35 patients) received placebo along with a daily insulin regimen for 3 months. There were no statistically significant differences in age, gender, duration of diabetes, body mass index, and blood pressure between the two groups. The authors observed significant reduction in glycosylated hemoglobin levels in patients receiving biotin supplement (HbA_{1c} 9.84 ± 1.80 at base, and after 3 months of treatment, it declined to 8.88 ± 1.73). However, in the control group, glycosylated hemoglobin concentrations were increased (HbA_{1c}, 9.39 ± 1.58 at the baseline, and after 3 months, it increased to 10.11 ± 1.68). There were statistically significant differences in the mean values of HbA_{1c} in both the biotin and the control groups ($P < .001$). The fasting blood sugar concentrations were significantly reduced in patients taking biotin (275 ± 65.76 mg/dL at the baseline and 226 ± 41.31 mg/dL after 3 months) but did not change in the control group (274.85 ± 74.40 mg/dL at the baseline and 276.47 ± 50.67 mg/dL after 3 months). There were also statistically significant differences in the means of total cholesterol, low-density lipoprotein cholesterol, and triglyceride between the two groups at the end of 3 months. The authors concluded that biotin administration as an adjuvant to insulin regimen can improve glycemic management and decrease plasma lipid concentrations in poorly controlled type 1 diabetic patients.[39]

Biotin Supplement and Hypertension

The pharmacologic effects of biotin on hypertension in the spontaneously hypertensive stroke-prone (SHRSP) rat strain have been investigated. In one study

the authors observed that long-term administration of biotin decreased systolic blood pressure in the SHRSP rat strain. Moreover, a single dose of biotin also immediately decreased systolic blood pressure in this strain. Biotin also reduced coronary arterial thickening and the incidence of stroke in the SHRSP rat strain. The authors speculated that pharmacologic doses of biotin decreased the blood pressure in the SHRSP rat via nitric oxide–independent direct activation of soluble guanylate cyclase. Therefore biotin may have beneficial effects on hypertension and may reduce the risk of stroke.[40]

A South African study involving children aged 10–15 years (321 black males and 373 females) from rural to urbanized communities, of which 40 male and 79 female subjects were identified with high-normal to hypertensive blood pressure, the authors using a stepwise regression analysis observed that lower intake of biotin, folic acid, pantothenic acid, zinc, and magnesium was significantly associated with blood pressure parameters of hypertensive males, whereas lower intake of biotin and vitamin A was significantly associated with blood pressure parameters of hypertensive females. The authors concluded that the dietary results coupled with the cardiovascular parameters of this study identified folic acid and biotin as the risk markers that could contribute to the cause of hypertension in black persons. Low intake of these nutrients, among others, is a matter of serious concern because this may increase the risk of hypertension.[41]

Biotin and Immune Functions

Studies with mice indicate that biotin deficiency enhances inflammation and adversely affects immune functions. Moreover, nutritional biotin deficiency and genetic defects in either holocarboxylase synthetase or biotinidase in humans induce cutaneous inflammation and immunologic disorders. Transcriptional factors, including nuclear factor κB and transcriptional factors Sp1/3, are affected by the status of biotin, indicating that biotin regulates immunologic and inflammatory functions independent of the biotin-dependent carboxylases.[42] Biotin affects the functions of adaptive immune T and natural killer cells, and biotin deficiency enhances the inflammatory response of human dendritic cells.[43]

Using five healthy adult volunteers who consumed 0.75 mg of biotin supplement per day for 14 days, Zempleni et al. showed that such supplementation resulted in a significant decrease in mitogen-stimulated peripheral blood mononuclear cell (PBMC) proliferation. Moreover, secretion of cytokines such as interleukin (IL)-1β and IL-2 decreased by 65% and 45%, respectively, following biotin supplementation. The authors concluded that administration of biotin for 14 days decreases PBMC proliferation and synthesis of IL-1β and IL-2. Therefore biotin has an inhibitory effect on PBMC proliferation and cytokine release..[44] Using six healthy adult subjects (one man and five women; age, 2–25 years) who took 2.15 mg of biotin daily for 5 days, Wiedmann observed that biotin supplementation increased expression of both the genes encoding the cytokines IL-1β, interferon-γ, and IL-4 in human PBMCs and the genes encoding 3-methylcrotonyl-CoA carboxylase (α-chain). The increased expression of the gene encoding 3-methylcrotonyl-CoA carboxylase may be associated with enhanced immune function. However, increased expression of the genes encoding the cytokines such as IL-1β and interferon-γ is not necessarily paralleled by increased levels of these cytokines in extracellular fluids. Biotin supplementation causes increased synthesis of cytokine receptors, leading to increased endocytosis and, thus, decreased extracellular accumulation of cytokines.[45]

Biotin deficiency in humans may increase the risk of infection due to compromised immune system. Candida infections secondary to impaired immune function might also contribute to the dermatitis due to biotin deficiency in infants and children.[46]

Case Report: A 38-year-old white woman presented with a 14-month history of persistent vaginal candidiasis despite therapy with fluconazole, ketoconazole, miconazole, gentian violet, and acidophilus tablets. She has two children with partial biotinidase deficiency, but her biotinidase enzyme activity, and that of her husband, put them in the carrier range. She also experienced episodes of recurrent bacterial vaginosis treated with metronidazole and vaginal douching with vinegar. The patient, knowing her carrier status for biotinidase deficiency, called her geneticist for advice and then started taking 20 mg biotin supplement daily. This dose is more than 60 times the minimum daily requirement, and crystalline biotin (Roche Pharmaceuticals, Nutley, NJ) must be ordered by a pharmacist and either weighed individually into capsules or measured by the patient with a small scoop. Within 2 months, her vulvovaginal symptoms disappeared completely. She has been symptom-free for 5 months, with no other therapy, at the time this report was submitted. She will continue on 20 mg of daily biotin indefinitely. The authors suggest given that 1 in every 123 individuals is predicted to be a carrier of biotinidase deficiency, there might be other women with chronic vaginal candidiasis who will respond to oral biotin therapy.[47]

Other Applications of Biotin Therapy

Although hemodialysis has the potential to deplete biotin, in one study the authors observed higher serum biotin levels (0.5–2.4 ng/mL, with one value of 3 ng/mL, n = 88) than those of healthy subjects (0.1–0.8 ng/mL, with one value 0.9 ng/mL, n = 51). Interestingly, 75 uremic patients on hemodialysis (85%) showed serum biotin levels above the upper limit of normal serum biotin levels.[48] Jung et al.[49] commented that regular biotin supplementation is not required in uremic patients.

Encephalopathy and peripheral neuropathy commonly develop in uremic patients. In one study the authors reported that pharmacologic dose of biotin is effective in alleviating neurologic symptoms in uremic patients. Yatzidis et al. reported that nine patients undergoing chronic hemodialysis for 2–10 years and suffering from encephalopathy (dialysis dementia) and peripheral neuropathy showed marked improvement within 3 months following 10 mg of biotin (given daily in three doses) therapy. Such improvements regarding symptoms of disorientation, speech disorders, memory failure, myoclonic jerks, flapping tremor, restless legs, paresthesia, and difficulties in walking were maintained for 15–25 months during follow-up in six patients, while the other three died of renal failure. The authors recommend initiating biotin therapy regularly in any patient with advanced renal failure before manifestation of severe neural or muscular lesions.[50] Another study also reported beneficial effect of biotin (10 mg/day) in treating neurologic disorders of uremia.[51]

A diet that is marginally deficient in biotin may cause sudden unexpected death of young broiler chickens when they are exposed to stress. Johnson et al. investigated whether biotin deficiency may cause sudden death in human infants by measuring biotin levels using a radiochemical technique in 204 livers obtained from infants at autopsy. The levels of biotin in the livers of infants who had died of sudden infant death syndrome (SIDS) were significantly lower than those in livers of infants of similar age, who had died of known causes. The authors concluded that their study supports an association between biotin and SIDS.[52]

Loss of taste is a very distressing condition for anyone. Greenway et al. reported two cases in whom loss of taste responded to oral biotin therapy. The first patient was a 67-year-old women who lost her sense of taste after taking two pills of Juvenon, a dietary supplement that delivers a daily total of 1 g of L-carnitine, 400 mg α-lipoic acid, 300 μg biotin, 154 mg calcium, and 117 mg phosphorus. She did not respond to 5 mg per day oral biotin therapy for 4 days but her taste

TABLE 3.3
Effects of Biotin in Human Health

Intended Use of Biotin Supplement	Comments
Healthy hair	Although biotin deficiency may cause hair loss, there is no scientific evidence supporting that biotin supplement helps hair growth in healthy individuals without any biotin deficiency.
Healthy nails	Biotin supplement of 2.5 mg/day is effective in treating brittle fingernails.
Healthy skin	Although biotin supplement may be effective in treating skin rashes caused by biotin deficiency in children, there is inconclusive evidence that biotin supplement may improve skin condition of healthy individuals with no biotin deficiency.
Treating symptoms of multiple sclerosis	Preliminary data from clinical trials indicate that high-dosage biotin therapy (300 mg/day) reduces symptoms of multiple sclerosis in some patients. No biotin toxicity was observed in any patient.
Lowering fasting glucose in diabetic patients	In one study, 9 mg biotin supplementation significantly reduces fasting glucose levels in patients with type 2 diabetes. Other studies have shown the glucose-lowering effect of biotin supplement in both type 1 and type 2 diabetic patients.
Lowering lipid levels	Biotin supplement may be useful to lower serum triglyceride and possibly cholesterol concentration.
Reducing hypertension	Biotin supplement reduces blood pressure in spontaneously hypertensive stroke-prone rats. A South African study indicates that lower intake of biotin was associated with hypertension in black children.
Immune function	Biotin has an immunomodulatory effect because its deficiency may increase risk of infection. There is a case report of resolution of chronic vaginal candidiasis with oral biotin therapy in a woman who is a carrier of biotinidase deficiency.
Basal ganglia disease	Biotin-responsive basal ganglia disease is a rare disease that responds to biotin therapy (see Chapter 2).
Other applications	Oral 10 mg biotin supplement may be effective in reducing symptoms of encephalopathy and peripheral neuropathy that commonly develop in uremic patients. Loss of taste may also be restored with pharmacologic dosage of biotin.

returned to 90% normal after using 10 mg biotin per day (5 mg biotin twice daily) for 3 days. She continued on the 5-mg twice daily dose, and after 1–2 weeks, her taste fully returned to normal. After 3 months of taking biotin, her taste remained normal. The second patient, a 60-year-old man, lost his taste after a sleeve gastrectomy for obesity. He did not have restoration of taste after taking 5 mg biotin supplement every day for 7 weeks but did have taste restoration on 20 mg per day biotin (5 mg supplement four times a day) therapy within 36 h. He then reduced the biotin dosage to 5 mg twice a day and his taste became metallic again. Upon increasing the dosage to 20 mg per day, his taste gradually returned to normal over a few days and remained so for 1 month. Interestingly, both subjects showed no indication of biotinidase deficiency and neither subjects had an abnormal biotinidase level. Therefore the effect of biotin in these two cases must represent

a pharmacologic effect of biotin rather than correcting biotin deficiency. The authors commented that regardless of the reasons of restoration of taste with oral biotin therapy, their observation is clinically important because loss of taste is very distressing to patients and impairs their quality of life significantly. The authors concluded that as biotin up to 40 mg per day has been shown to be safe, therapeutic trials of pharmacologic doses of biotin should be considered as a potentially curative therapy in patients who present with a loss of taste for no obvious reason.[53] Effects of biotin in human health are listed in Table 3.3.

Potential Toxicity of Biotin Supplements

Biotin is a water-soluble vitamin and is not considered toxic. Potential toxicity of biotin is addressed in Chapter 2. Briefly, in one study, the authors observed no adverse effect when biotin as a dosage of 5 mg/day was administered for up to 2 years.[30] Another report showed no toxicity of oral biotin at a dosage of 10 mg/day for 1 year in a 15-year-old boy treated for biotinidase deficiency.[54] Transient diarrhea is the only adverse effect reported during high-dose biotin therapy in patients suffering from multiple sclerosis.[22] Biotin is used as a hair-conditioning agent and a skin-conditioning agent in many cosmetic products at concentrations ranging from 0.0001% to 0.6%. Only minor acute oral toxicity has been reported in animal tests but biotin was not mutagenic in bacterial tests; although there is one case study in the literature reporting an urticarial reaction, a very large number of individuals exposed to biotin on a daily basis reported no adverse effects. The authors concluded that use of biotin is safe in cosmetic products.[55] However, in one study the authors observed that biotin supplementation affects liver morphology in normal mice, but these modifications do not alter biomarkers of liver damage.[56] Biotin supplementation affects the expression of numerous genes. However, it is unknown whether any of these alterations are undesirable.[37]

CONCLUSIONS

Daily biotin requirement is relatively low and can be satisfied by eating a balanced diet. Multivitamins also contain sufficient biotin (30–60 μg) to satisfy the daily requirement of 30 μg. Nevertheless, people take biotin supplements because there is a popular belief that taking biotin supplements results in hair and nail growth, as well as healthy skin. However, scientific research fails to confirm such popular claims because biotin supplements improve hair growth only in individuals who are deficient in biotin but not in healthy individuals without any biotin deficiency. However, scientific research indicates that 2.5 mg daily biotin supplement is effective in restoring brittle nails.

Scientific research indicates that biotin supplement may improve glucose control in diabetic patients. Biotin may also be effective in improving lipid profile. Limited data also indicates that when biotin was given in pharmaceutical dosage (100–300 mg/per day), it may be effective in alleviating symptoms in a subset of patients suffering from multiple sclerosis. Biotin is a water-soluble vitamin and is safe. However, taking biotin in a daily dosage exceeding 5 mg may interfere with clinical laboratory test results in biotin-based immunoassays. Biotin falsely increases analyte values in competitive immunoassays but falsely lowers analytical values in sandwich assay format. Wrong diagnosis of hyperthyroidism due to biotin use has been well documented in the medical literature. An FDA report indicates death of a person due to missed diagnosis of myocardial infarction because of falsely lower troponin value reported in a person who took biotin supplements. This important issue is discussed in detail in Chapter 4.

REFERENCES

1. Bonjour JP. Biotin in human nutrition. *Ann N Y Acad Sci.* 1985;447:97–104.
2. Biotin fact sheets for health professionals. National Institute of Health: office of Dietary Supplement.
3. Samarsinghe S, Meah F, Singh V, Basit A, et al. Biotin interference with routine clinical immunoassays: understanding the causes and mitigate risk. *Endocr Pract.* 2017;23:989–998.
4. Patel DP, Swink SM, Castelo-Soccio L. A review of the use of biotin for hair loss. *Skin Appendage Disord.* 2017;3:166–169.
5. Limat A, Suormala T, Hunziker T, Waelti ER, et al. Proliferation and differentiation of cultured human follicular keratinocytes are not influenced by biotin. *Arch Dermatol Res.* 1996;288:31–38.
6. Soleymani T, Lo Sicco K, Shapiro J. The infatuation with biotin supplementation: is there truth behind its rising popularity? A comparative analysis of clinical efficacy versus social popularity. *J Drugs Dermatol.* 2017;16:496–500.
7. Trüeb RM. Serum biotin levels in women complaining of hair loss. *Int J Trichol.* 2016;8:73–77.
8. Swamy SS, Ravikumar BC, Vinay KN, Yashovardhana DP, et al. Uncombable hair syndrome with a woolly hair nevus. *Indian J Dermatol Venereol Leprol.* 2017;83:87–88.
9. Shelley WB, Shelley ED. Uncombable hair syndrome: observations on response to biotin and occurrence in siblings with ectodermal dysplasia. *J Am Acad Dermatol.* 1985;13:97–102.
10. Boccaletti V, Zendri E, Giordano G, Gnetti L, et al. Familial uncombable hair syndrome: ultrastructural hair study and response to biotin. *Pediatr Dermatol.* 2007;24:E14–E16.
11. Grootens KP, Hartong EGTM. A case report of biotin treatment for valproate-induced hair loss. *J Clin Psychiat.* 2017;78:e838.
12. Mercke Y, Sheng H, Khan T, Lippmann S. Hair loss in psychopharmacology. *Ann Clin Psychiat.* 2000;12:35–42.
13. Ramakrishnappa SK, Belhekar MN. Serum drug level-related sodium valproate-induced hair loss. *Indian J Pharmacol.* 2013;45:187–188.
14. Colombo VE, Gerber F, Bronhofer M, Floersheim GL. Treatment of brittle fingernails and onychoschizia with biotin: scanning electron microscopy. *J Am Acad Dermatol.* 1990;23:1127–1132.
15. Hochman LG, Scher RK, Meyerson MS. Brittle nails: response to daily biotin supplementation. *Cutis.* 1993;51(4):303–305.
16. Floersheim GL. Treatment of brittle fingernails with biotin. *Z Hautkr.* 1989;64:41–48. [Article in German].

17. Piraccini BM, Tullo S, Iorizzo A, Rech G, et al. Triangular worn-down nails: report of 14 cases. *G Ital Dermatol Venereol.* 2005;140:136–140.
18. Möhrenschlager M, Schmidt T, Ring J, Abeck D. Recalcitrant trachyonychia of childhood - response to daily oral biotin supplementation: report of two cases. *J Dermatolog Treat.* 2000;11:113–115.
19. Lipner SR, Scher RK. Biotin for treatment of nail diseases: what is the evidence? *J Dermatolog Treat.* 2018;29:411–414.
20. Lipner SR. Rethinking biotin therapy for hair, nail and skin disorders. *J Am Acad Dermatol.* 2018;78:1236–1238.
21. Sedel F, Bernard D, Mock DM, Tourbah A. Targeting demyelination and virtual hypoxia with high-dose biotin as a treatment for progressive multiple sclerosis. *Neuropharmacology.* 2016;110(Pt B):644–653.
22. Sedel F, Papeix C, Bellanger A, Touitou V, et al. High doses of biotin in chronic progressive multiple sclerosis: a pilot study. *Mult Scler Relat Disord.* 2015;4:159–169.
23. Tourbah A, Lebrun-Frenay C, Edan G, Clanet M, et al. MD1003 (high-dose biotin) for the treatment of progressive multiple sclerosis: a randomized, double-blind, placebo-controlled study. *Mult Scler.* 2016;22:1719–1731.
24. Trambas CM, Sikaris KA, Lu ZX. A caution regarding high-dose biotin therapy: misdiagnosis of hyperthyroidism in euthyroid patients. *Med J Aust.* 2016;205:192.
25. Zhang H, Osada K, Maebashi M, Ito M, et al. A high biotin diet improves the impaired glucose tolerance of long-term spontaneously hyperglycemic rats with non-insulin-dependent diabetes mellitus. *J Nutr Sci Vitaminol.* 1996;42:517–526.
26. McCarty MF. cGMP may have trophic effects on beta cell function comparable to those of cAMP, implying a role for high-dose biotin in prevention/treatment of diabetes. *Med Hypothesis.* 2006;66:323–328.
27. Maebashi M, Makino Y, Furukawa Y. Ohinata K eta l. Therapeutic evaluation of the effects of biotin on hyperglycemia in patients with non-insulin dependent diabetes mellitus. *J Clin Biochem Nutr.* 1993;14:211–218.
28. Kouzi SA, Yang S, Nuzum DS, Dirks-Naylor AJ. Natural supplements for improving insulin sensitivity and glucose uptake in skeletal muscle. *Front Biosci (Elite Ed).* 2015;7:94–106.
29. Coggeshall JC, Heggers JP, Robson MC, Baker H. Biotin status and plasma glucose in diabetes. *Ann N.Y Acad Sci.* 1985;447:389–392.
30. Koutsikos D, Agroyannis B, Tzanatos-Exarchou H. Biotin for diabetic peripheral neuropathy. *Biomed Pharmacother.* 1990;44:511–514.
31. Singer GM, Geohas J. The effect of chromium picolinate and biotin supplementation on glycemic control in poorly controlled patients with type 2 diabetes mellitus: a placebo-controlled, double-blinded, randomized trial. *Diabetes Technol Ther.* 2006;8:636–643.
32. Báez-Saldaña A, Zendejas-Ruiz I, Revilla-Monsalve C, Islas-Andrade S, et al. Effects of biotin on pyruvate carboxylase, acetyl-CoA carboxylase, propionyl-CoA carboxylase, and markers for glucose and lipid homeostasis in type 2 diabetic patients and nondiabetic subjects. *Am J Clin Nutr.* 2004;79:238–243.
33. Lazo de la Vega-Monroy ML, Larrieta E, German MS, Baez-Saldana A, et al. Effects of biotin supplementation in the diet on insulin secretion, islet gene expression, glucose homeostasis and beta-cell proportion. *J Nutr Biochem.* 2013;24:169–177.
34. Aguilera-Méndez A, Fernández-Mejía C. The hypotriglyceridemic effect of biotin supplementation involves increased levels of cGMP and AMPK activation. *Biofactors.* 2012;38:387–394.
35. Larrieta E, Velasco F, Vital P, López-Aceves T, Lazo-de-la-Vega-Monroy ML, et al. Pharmacological concentrations of biotin reduce serum triglycerides and the expression of lipogenic genes. *Eur J Pharmacol.* 2010;644:263–268.
36. Marshall MW, Kliman PG, Washington VA, Mackin JF, et al. Effects of biotin on lipids and other constituents of plasma of healthy men and women. *Artery.* 1980;7:330–351.
37. Revilla-Monsalve C, Zendejas-Ruiz I, Islas-Andrade S, Báez-Saldaña A, et al. Biotin supplementation reduces plasma triacylglycerol and VLDL in type 2 diabetic patients and in nondiabetic subjects with hypertriglyceridemia. *Biomed Pharmacother.* 2006;60:182–185.
38. Dokusova OK, Krivoruchenko IV. The effect of biotin on the level of cholesterol in the blood of patients with atherosclerosis and essential hyperlipidemia. *Kardiologiia.* 1972;12:113. [Article in Russian].
39. Hemmati M, Babaei H, Abdolsalehei M. Survey of the effect of biotin on glycemic control and plasma lipid concentrations in type 1 diabetic patients in kermanshah in Iran (2008–2009). *Oman Med J.* 2013;28:195–198.
40. Watanabe-Kamiyama M, Kamiyama S, Horiuchi K, Ohinata K, et al. Antihypertensive effect of biotin in stroke-prone spontaneously hypertensive rats. *Br J Nutr.* 2008;99:756–763.
41. Schutte AE, van Rooyen JM, Huisman HW, Kruger HS, et al. Dietary risk markers that contribute to the aetiology of hypertension in black South African children: the THUSA BANA study. *J Hum Hypertens.* 2003;17:29–35.
42. Kuroishi T. Regulation of immunological and inflammatory functions by biotin. *Can J Physiol Pharmacol.* 2015;93:1091–1106.
43. Agrawal S, Agrawal A, Said HM. Biotin deficiency enhances the inflammatory response of human dendritic cells. *Am J Physiol Cell Physiol.* 2016;311:C386–C391.
44. Zempleni J, Helm RM, Mock DM. In vivo biotin supplementation at a pharmacologic dose decreases proliferation rates of human peripheral blood mononuclear cells and cytokine release. *J Nutr.* 2001;131:1479–1484.
45. S W, Eudy JD, Zempleni J. Biotin supplementation increases expression of genes encoding interferon-gamma, interleukin-1beta, and 3-methylcrotonyl-CoA carboxylase, and decreases expression of the gene encoding interleukin-4 in human peripheral blood mononuclear cells. *J Nutr.* 2003;133:716–719.
46. Mock DM. Skin manifestations of biotin deficiency. *Semin Dermatol.* 1991;10:296–302.
47. Strom CM, Levine EM. Chronic vaginal candidiasis responsive to biotin therapy in a Carrier of biotinidase deficiency. *Obstet Gynecol.* 1998;92:644–646.

48. Livaniou E, Evangelatos GP, Ithakissios DS, Yatzidis H, et al. Serum biotin levels in patients undergoing chronic hemodialysis. *Nephron.* 1987;46:331–332.

49. Jung U, Helbich-Endermann M, Bitsch R, Schneider S, et al. Are patients with chronic renal failure (CRF) deficient in Biotin and is regular Biotin supplementation required? *Z Ernahrungswiss.* 1998;37:363–367.

50. Yatzidis H, Koutsicos D, Agroyannis B, Papastephanidis C, et al. Biotin in the management of uremic neurologic disorders. *Nephron.* 1984;36:183–186.

51. Yatzidis H, Koutsicos D, Alaveras AG, Papastephanidis C, et al. Biotin for neurologic disorders of uremia. *N Engl J Med.* 1981;305:764.

52. Johnson AR, Hood RL, Emery JL. Biotin and the sudden infant death syndrome. *Nature.* 1980;285(5761):159–160.

53. Greenway FL, Ingram DK, Ravussin E, Hausmann M, et al. Loss of taste responds to high-dose biotin treatment. *J Am Coll Nutr.* 2011;30:178–181.

54. Ramaekers VT, Brab M, Rau G, Heimann G. Recovery from neurological deficits following biotin treatment in a biotinidase Km variant. *Neuropediatrics.* 1993;24:98–102.

55. Fiume MZ. Cosmetic ingredient review expert panel. Final report on the safety assessment of biotin. *Int J Toxicol.* 2001;20(suppl 4):1–12.

56. Riverón-Negrete L, Sicilia-Argumedo G, Álvarez-Delgado C, Coballase-Urrutia E, et al. Dietary biotin supplementation modifies hepatic morphology without changes in liver toxicity markers. *BioMed Res Int.* 2016;2016:7276463.

Effect of Biotin on Clinical Laboratory Test Results: How to Avoid Such Interferences?

INTRODUCTION

Biotin (vitamin B$_7$) previously known as vitamin H is an essential vitamin that acts as a coenzyme for carboxylase reactions in gluconeogenesis, fatty acid synthesis, and amino acid catabolism. Biotin is present in many foods including some vegetables, fruits (banana, for example), salmon, chicken, eggs, whole-grain cereals, whole wheat bread, dairy products, and nuts. The daily recommended allowance of biotin is only 30 µg. The average dietary biotin intake in Western population has been estimated to be 35–70 µg/day. For this reason, statutory agencies in some countries do not prescribe a recommended daily intake of biotin.[1,2] Biotin deficiency is rare, although pregnant women may be susceptible to biotin deficiency. Moreover, genetic predisposition may cause biotin deficiency. Please see Chapter 2 for a more detailed discussion.

According to the Academy of Nutrition and Dietetics, consuming a variety of nutrient-dense foods and beverages in moderation can provide adequate calories, dietary fibers, macronutrients, and micronutrients for adults and children. Although eating a balanced diet every day is sufficient to fulfill all requirements of needed vitamins and minerals, vitamin and mineral supplements are the largest growing category of dietary supplements consumed by people worldwide.[3] Use of vitamin and mineral supplements is very popular among men and women over 50 years of age, making the annual sales of these supplements over $11 billion, although scientific research indicates that for the majority of population, there is no overall health benefit from taking vitamin and mineral supplements.[4]

Biotin is included in most multivitamin supplements but usually in the lower dosages. Many automated immunoassays incorporate biotinylated antibodies and streptavidin-coated magnetic beads as a means of immobilizing antigen-antibody complexes to the solid phase. This design offers many advantages, including signal amplification and increased sensitivity. Biotin-based immunoassays are used in clinical laboratories

for decades. However, there have been reports of clinically significant interferences of biotin in immunoassays using biotinylated antibodies. This coincides with the tendency by people to use high-dose biotin supplements because serum biotin levels in people taking no supplement as well as multivitamin supplements (usually contains 30 µg biotin) are usually low and not sufficient to cause any clinically significant interference. However, biotin is available at moderate to high doses starting from 500-µg to 10-mg capsules. Over-the-counter formulations containing 100 mg biotin capsules are also available. People take biotin supplements for hair and nail growth as well as for a healthier skin (please see Chapter 3). Moreover, patients with biotin deficiency may receive biotin supplementation of 40–100 mg per day as a part of therapy. In addition, very high dosages of biotin (300 mg or higher daily) may be used for treating multiple sclerosis and demyelinating conditions.[1] Clinically significant interference in some biotin-based immunoassays occurs when people take large doses of biotin supplement per day.[5]

Biotin supplements, particularly at large doses, have become increasingly popular as approximately 15%–20% people in the United States consume biotin-containing supplements. Biotin is water soluble and very safe. There are favorable tolerability and safety profiles in individuals who receive pharmacologic dosages up to 300 mg per day.[6] Retail sales of biotin supplements grew more than 260% between 2013 and 2016. The total sales are probably much higher because the Nelson Food Drug sales data do not include sales from wholesale warehouses, vitamin and beauty product retail stores, and online sales from sites where biotin is the top selling product.[7]

THE FOOD AND DRUG ADMINISTRATION SAFETY COMMUNICATION ON BIOTIN

The Food and Drug Administration (FDA) posted a safety communication, alerting the public, healthcare

Biotin and Other Interferences in Immunoassays. https://doi.org/10.1016/B978-0-12-816429-7.00004-6

providers, laboratory personnel, and laboratory test developers that biotin can significantly interfere with certain laboratory test results, which may go undetected in patients who are ingesting high levels of biotin as dietary supplements. The FDA has seen an increase in the number of reported adverse events, including one death related to biotin interference with laboratory test results. The FDA has received a report that one patient taking high levels of biotin died from falsely low troponin results when a troponin test known to have biotin interference was used. However, the FDA did not specify which diagnostic company marketed the troponin assay.[8]

The FDA is aware of people taking high levels of biotin that would interfere with laboratory test results. Certain dietary supplements promoted for healthy hair, nails, and skin contain biotin levels 650 times the recommended daily intake of biotin. Physicians may treat certain conditions such as multiple sclerosis using high doses of biotin. The FDA has the following recommendations to healthcare providers:

- Ask patients whether they are taking any dietary supplements including supplements marketed for hair, skin, and nail growth.
- Be aware that many laboratory tests including but not limited to cardiovascular diagnostics tests and hormone tests that use biotin-based technology may be affected and incorrect result may be generated if biotin is present in patient's specimen.
- Communicate with laboratory personnel if a patient is taking biotin supplement.
- If the laboratory test result does not match with clinical presentation of a patient, biotin interference should be considered as a source of error in the laboratory test results.
- Healthcare providers must be aware that biotin found in multivitamins, prenatal multivitamins, biotin supplements, and products promoted for healthy skin and hair and nail growth may contain enough biotin to cause interferences with laboratory tests. Healthcare providers should report any adverse patient outcome due to biotin interference to the FDA and manufacturer of the laboratory test.

The FDA also issued the following recommendations for laboratory personnel:

- If biotin-based immunoassays are used in the laboratory, it is difficult to identify specimens containing biotin; therefore, it important to communicate with healthcare providers and patients in order to prevent biotin interferences in laboratory test results. It is important to educate healthcare providers about biotin interferences in certain laboratory test results.
- If specimens are collected in the laboratory then it is important to ask patients whether they are taking biotin.
- Although taking the daily recommended allowance of biotin ($30\,\mu g$) usually has no effect on biotin-based immunoassays, some biotin supplements may contain $20\,mg$ of biotin and a biotin level of $300\,mg$ may also be used for therapy in patients with multiple sclerosis. Biotin levels higher than the recommended daily allowance may significantly affect laboratory test results.
- Laboratory personnel should be aware that specimens collected from patients taking high amounts of biotin may produce serum biotin levels more than $100\,ng/mL$. Concentration of biotin up to $1200\,ng/mL$ may be present in specimens collected from patients taking $300\,mg$ of biotin daily as a part of therapy.
- Currently there is insufficient evidence to support recommendations for safe testing in patients taking high levels of biotin, including about the length of time required for clearance of biotin from blood. Laboratory personnel should communicate with assay manufacturers about biotin interference.

The FDA also has the following recommendations for consumers:

- Talk to your doctor if you are currently taking biotin or are considering adding biotin, or a supplement containing biotin, to your diet.
- Know that biotin is found in multivitamins, including prenatal multivitamins, biotin supplements, and supplements for hair, skin, and nail growth, in levels that may interfere with laboratory tests. The FDA urges the general public to know that biotin is found in many over-the-counter supplements in levels that may interfere with laboratory tests. Examples are B-complex vitamins; coenzyme R; dietary supplements for hair, skin, and nail growth; multivitamins; prenatal vitamins; and vitamin B_7 and vitamin H supplements (biotin is also known as vitamin B_7 and vitamin H).
- Be aware that some supplements, particularly those labeled for hair, skin, and nail benefits, may have high levels of biotin, which may not be clear from the name of the supplement.
- If you have had a laboratory test done and are concerned about the results, then talk to your healthcare provider about the possibility of biotin interference.

SERUM BIOTIN LEVELS

As clearly mentioned in the FDA safety communication, consuming biotin as recommended by daily allowance should not have any significant effect on laboratory test results. This is the reason why there is no report of biotin interference in immunoassays using biotinylated antibody in the old medical literature when people took no biotin supplement or consumed multivitamin tablets containing low amounts of biotin. Moreover, taking biotin in a standard dosage of 500 μg–1 mg daily may not cause any significant interference with biotin-based immunoassays because serum biotin levels are usually well below the threshold biotin concentrations needed for clinically significant interference. However, people are consuming higher amounts of biotin as supplements that may cause significant interference in biotin-based immunoassays. In one review the authors commented that biotin interference in clinical laboratory test results is a relatively new discovery.[9] The interference of biotin in biotin-based immunoassays is correlated with biotin dose and serum biotin concentrations after taking such supplements. Therefore, in the following section, serum biotin levels after taking various doses of biotin supplement are discussed.

In general, biotin concentrations are very low in people not taking any supplement. The biotin deficiency only occurs at a serum level of <0.2 ng/mL. Suboptimal level of biotin is 0.2–0.4 ng/mL, while biotin level above 0.4 ng/mL is considered adequate. For subjects not taking supplement, biotin concentrations usually vary from 0.4 to 1.2 ng/mL.[1] In another report, blood concentration of biotin in healthy individuals not taking supplements ranged from 0.12 to 0.36 nmol/L.[10] Based on a study of 51 subjects, Livaniou et al. reported that, in general, biotin levels varied from 0.1 to 0.8 ng/mL in healthy men and women, although one subject showed a biotin level of 0.9 ng/mL. Interestingly, serum biotin levels were higher (0.5–2.4 ng/mL) in patients undergoing chronic hemodialysis for which authors had no explanation.[11] Watanabe et al.[12] reported that the mean biotin level in normal healthy Japanese population was 2.3 ng/mL but the mean level in elderly Japanese people aged 65 years and more was 2.5 ng/mL. The observed normal level of biotin is higher in the Japanese population than the reported normal biotin levels in other population.

However, taking biotin supplement results in a higher serum biotin level, which is dose dependent. Based on a study of 12 men (mean age, 44 years) and 10 women (mean age, 46 years), Clevidence et al. observed that mean plasma biotin level in both men and women not taking any biotin supplements varied from 0.41 to 0.66 ng/mL. However, biotin levels were significantly higher in men and women after taking biotin

supplements. Plasma and urine biotin levels were highest 2 h after taking biotin supplement, decreased between 2 and 3 h, and then were reduced to almost baseline levels after 24 h. For example, after taking 75 μg of biotin supplement, average biotin concentration in plasma was increased to 0.89 ng/mL 2 h after ingestion of biotin, then declined to 0.75 ng/mL in 4 h, and finally reduced to 0.61 ng/mL in 24 h. However, biotin levels were increased to a higher magnitude in women after taking biotin supplement—in women, mean biotin concentration in plasma was increased to 1.38 ng/mL 2 h after taking 75 μg supplement. The mean value was reduced to 0.60 ng/mL after 24 h. As expected, biotin levels were higher in both men and women after ingesting 150 μg of biotin supplement than those when they took 75 μg supplement. However, plasma biotin levels were increased approximately fourfold in men and sixfold in women after consuming 300 μg of biotin supplement. In men, plasma biotin levels increased from 0.59 to 2.77 ng/mL 2 h after taking 300 μg biotin supplement and then declined to 0.84 ng/mL after 24 h. In women, 2 h after receiving 300 μg of biotin the mean plasma biotin level increased from 0.42 to 3.11 ng/mL. Up to 33% of the ingested biotin was excreted in urine within 4 h.[13] In another study involving 15 volunteers the average biotin plasma concentration was 3.7 ng/mL (range, 2.0–5.8 ng/mL) 3 hours after taking 1.2 mg (1200 μg) biotin supplement acutely, but after taking 1.2 mg biotin supplement daily for 2 weeks the average plasma biotin concentration was 5.5 ng/mL (2.3–11.6 ng/mL).[14] Extrapolating this data, it is expected that biotin concentration in plasma should be around 15 ng/mL after ingesting 5 mg of biotin supplement and 30 ng/mL after ingesting 10 mg of biotin supplement.[15]

Mardach et al.[16] reported that when five normal adults consumed 1 mg of biotin daily, the mean biotin level of 34,600 + 8000 pmol/L (mean biotin, 8.6 ng/mL) in blood was observed 1–3 h after the last biotin dose. Grimsey et al. studied serum biotin levels after volunteers consumed 5, 10, or 20 mg biotin supplement every day for 5 days. The authors observed that serum biotin concentrations were dose dependent. The median peak biotin concentration in serum (1 h after dose) on day 3 was 41 ng/mL (range, 10–73 ng/mL) after taking 5 mg biotin supplement each day. However, the median serum biotin concentration was 91 ng/mL (range, 53–141 ng/mL) after taking 10 mg supplement and 184 ng/mL (range, 80–355 ng/mL) after ingesting 20 mg biotin supplement each day. The half-life of biotin was calculated to be 15 h with consecutive dosing over several days.[17]

As expected, serum biotin levels would be much higher after taking biotin in pharmaceutical dosages such as 100 or 300 mg daily. The mean peak biotin concentration 1–3 hours after taking 100 mg biotin supplement was 495 ng/mL. The half-life of biotin is approximately 1.8 h after taking microgram amounts, but after intake of 300 mg biotin, half-life may be prolonged to 7.8–18.8 h. The plasma biotin concentrations varied from 700 to 900 ng/mL when subjects ingested a single dose of 300 mg biotin supplement and blood was drawn 2 h after taking the supplement. After 24 h the biotin plasma levels were reduced to around 75 ng/mL.[18]

Piketty et al. measured serum biotin levels along with the biotin metabolites bisnorbiotin and biotin sulfoxide in 27 specimens collected from 20 subjects receiving high dose of biotin (nine patients with multiple sclerosis receiving 300 mg biotin daily for 12 months; eight healthy controls receiving a single dose of 100, 200, or 300 mg biotin; and three authors volunteered to receive 15 or 30 mg of biotin). The lowest observed biotin concentration in plasma was 7 ng/mL in a subject 48 h after taking 15 mg biotin (but biotin level was 43.9 ng/mL in another volunteer who also consumed 15 mg of biotin but blood was drawn 2 h after taking the supplement) and the highest biotin concentration was 1160 ng/mL in a healthy subject 1 h after taking 300 mg biotin supplement. The lowest observed value in plasma was 669 ng/mL after taking 300 mg supplement and the specimen was collected after 1 hour. The plasma biotin concentrations varied from 169 ng/mL (collected 11 h 8 min after last dose) to 694 ng/mL (collected 105 min after the last dose) among patients taking 300 mg biotin supplement daily. The highest plasma biotin concentration was 1000 ng/mL in one healthy volunteer who consumed 200 mg of biotin supplement and the blood was collected after 2 h. The lowest plasma biotin level after taking 200 mg supplement was 758 ng/mL. The plasma biotin level in a volunteer was 407 ng/mL 2 h after taking 100 mg biotin supplement. After taking 30 mg biotin supplement, plasma biotin level was 56.8 ng/mL when blood was drawn after 2 h.[19] The FDA commented that specimen collected from patients taking 300 mg of biotin may reach up to 1200 ng/mL.[8] Serum or plasma levels of biotin in healthy population not taking biotin as well as in volunteers after taking various amounts of biotin supplements are listed in Table 4.1.

TABLE 4.1
Biotin Levels in Humans Taking No Supplement and After Taking Biotin Supplement

Biotin Supplement	Serum/Plasma Biotin Concentration (ng/mL)	References
No supplement	0.4–1.2	[1]
No supplement	0.1–0.8	[11]
No supplement	0.41–0.66	[13]
No supplement (Japanese population)	Mean: 2.3 Mean: 2.5 (>65 years of age)	[12]
300 μg	0.59–2.77 in men 2 h after taking supplement 0.42–3.11 in women 2 h after taking supplement	[13]
1.2 mg (1200 mg supplement)	2.3–11.6 (chronic intake; 1.2 mg per day for 2 weeks), 3 h after taking dose 2.0–5.8 (acute intake), 3 h after taking dose	[14]
1 mg supplement daily for 7 days	Mean level, 8.4, 1–3 h after the last biotin dose	[16]
5 mg	10–73, 1 h after taking dose	[17]
10 mg	53–141, 1 h after taking dose	[17]
20 mg	80–355, 1 h after taking dose	[17]
100 mg	Peak level, 495	[18]
300 mg	700–900, 2 h after taking dose	[18]
200 mg	758–1000, 2 h after taking dose	[19]
300 mg	669–1160, 1 h to 1 h 30 min after taking dose	[19]

INTERFERENCE OF BIOTIN IN IMMUNOASSAYS

Biotin is a small molecule that when conjugated with an active macromolecule, such as an assay antibody, rarely interferes with the function of the labeled molecule. Streptavidin is a glycoprotein that has a high affinity for biotin and the binding is not only highly specific but also resistant to changes in temperature, pH, and the presence of denaturing agents and organic solvents. By attaching streptavidin protein to a moiety such as a fluorophore, an enzyme, or a gold nanoparticle, the streptavidin-biotin interaction can be utilized as a detection method in a variety of immunoassays including enzyme-linked immunosorbent assay (ELISA), chemiluminescent assay (CLIA), as well as Western blotting and flow cytometry. ELISA and CLIA are routinely used in clinical laboratories to assay for hormones, vitamins, therapeutic drugs, tumor markers, and other targets.[20]

Streptavidin/biotin-based immunoassays are vulnerable to biotin interferences if biotin is present in excess concentration. Exogenous biotin present in the specimen competes with biotinylated reagents for binding sites on the streptavidin reagent. In competitive immunoassays, concentration of the analyte is inversely proportional to the signal intensity of the washed solid phase. Biotin reduces signal intensity in competitive immunoassays by occupying streptavidin-binding sites, thus falsely increasing the true analyte concentration. In sandwich immunoassay format the concentration of the analyte is directly proportional to the signal intensity of the washed solid phase. Exogenous biotin competes with the binding of labeled complex to the solid phase and reduces the signal intensity, thus producing false lower analyte concentration. A third immunoassay format that is susceptible to biotin interference utilizes biotinylated antigens and/or labels and an antibinding protein (other than streptavidin). Exogenous biotin competes with the biotinylated complex to bind with the antibiotin-coated solid phase, and depending

on the assay design, it produces positive interference in competitive immunoassays and negative interference in noncompetitive immunoassays. However, the mere presence of biotin/streptavidin in assay design does not necessarily imply that the assay will be subjected to biotin interference. In some assays the streptavidin (or antibiotin antibody) and the biotin-labeled reagents are combined during the manufacturing process (prebound reagents). Such assays are expected to be less susceptible or unaffected in the presence of high biotin concentrations in the serum or plasma.[6] In general, competitive immunoassays are used for detecting relatively small molecules, whereas sandwich assays are more effective in detecting large molecules. Please also see Chapter 1 for an in-depth discussion on immunoassay design and mechanism of biotin interference. As expected, only biotin-based immunoassays are affected by high concentrations of biotin. The list of automated analyzers that may use some biotin-based assays is presented in Table 4.2. The tests commonly affected due to biotin interferences are listed in Table 4.3.

Definition of Clinically Significant Interference

Clinical diagnostics manufacturers typically evaluate analytical interference by supplementing human serum or plasma with a potential interfering agent and then identifying a concentration at which the interfering substance causes a +10% change in the baseline value. Usually if the analytical bias is less than 10%, especially in the presence of a very higher concentration of the interfering substance, then it may be concluded that the analyte of interest is free from the interfering compounds. However, bias over 20% (positive or negative) is very problematic and is considered as clinically significant according to the criterion defined by the College of American Pathologists.[21] Interestingly, biotin is a unique interfering substance because it has no impact on clinical laboratory test results in patients taking no

TABLE 4.2
Automated Analyzers That Have Biotin-Based Immunoassays

Automated Analyzer	Diagnostic Company
Elecsys, cobas, and Modular platforms	Roche Diagnostics
Vitros platforms	Ortho Clinical Diagnostics
Dimension ExL, Dimension Vista, Immulite, Centaur platforms	Siemens Healthineers Laboratory Diagnostics
Access, DxI, and DXC platforms	Beckman Coulter
iSYS platform	Immunodiagnostic Systems

TABLE 4.3
Commonly Reported Biotin Interferences in Biotin-Based Immunoassays

Type of Test	Assays Showing Falsely Elevated Values (Competitive Immunoassays)	Assays Showing Falsely Lower Values (Sandwich Immunoassays)
Thyroid function test	Free T_4, total T_4, free T_3, total T_3, anti–thyroid peroxidase, anti–thyrotropin receptor antibody	Thyroid-stimulating hormone, thyroglobulin
Hormones	Testosterone, dehydroepiandrosterone sulfate, estradiol, progesterone, cortisol	Parathyroid hormone, luteinizing hormone, follicle-stimulating hormone, growth hormone, prolactin, insulin, C-peptide, adrenocorticotropin hormone
Cardiac markers		Troponin I, troponin T, creatine kinase MB
Serologic tests		HIV antigen/antibody combo, hepatitis C virus antibody, hepatitis B surface antigen, hepatitis B surface antibody, hepatitis B e antigen, hepatitis B core antibody IgM, hepatitis A antibody IgM, hepatitis A antibody total
Other analytes	25-Hydroxyvitamin D, vitamin B_{12}, folate	NT-proBNP, hCG, ferritin, myoglobin, sex hormone–binding globulin
Tumor marker		α-Fetoprotein, cancer antigen 125, carbohydrate antigen 19-9, PSA, free PSA
Therapeutic drug	Digoxin, digitoxin	

hCG, human chorionic gonadotropin; *NT-proBNP*, N-terminal prohormone brain natriuretic peptide; *PSA*, prostate-specific antigen.

supplement or multivitamins (with usually 30 μg biotin) or biotin supplements containing 500 μg to 1 mg biotin. Usually biotin supplement of 5 mg per day is needed to observe any significant assay interference. A biotin dose of 10 mg (more than 300 times the recommended daily allowance of 30 μg) may affect certain assays but not all biotin-based assays. However, if a person is taking a pharmaceutical dosage of biotin (100 or 300 mg per day), it will affect almost all biotin-based assays because biotin interference is very significant at 500 ng/mL or higher serum biotin concentrations, which can be easily achieved after taking 300 mg biotin supplement.

Although the first report of biotin interference in thyroid function tests was published in 1996, several case reports related to biotin interference in thyroid function tests emerged around 2012. As biotin half-life varies widely depending on the amount of biotin dose ingested, a particular patient may show different levels of interferences depending on several factors:

- How much biotin is ingested?
- How much time has elapsed between ingestion of biotin and time of blood draw?
- Which biotin-based immunoassay is used?

There are three types of studies to report clinically significant interference of biotin in biotin-based immunoassays:

- Case reports showing misdiagnosis due to biotin interference in clinical laboratory test results. Most case reports (children and adults) indicate wrong diagnosis of hyperthyroidism due to biotin interferences in thyroid function tests.
- Healthy volunteers taking known amount of biotin supplement and then predose and postdose analyte values were compared using biotin-based immunoassays. Maximum interferences were observed approximately 2 h after dose.
- In vitro studies where specimens were supplemented with a known amount of biotin and the results were compared before and after supplementation.

Although interferences in certain biotin-based assays have only been documented using in vitro studies, there are limitations to this approach. In general, in vitro supplementation studies can only evaluate interference of pure biotin in immunoassays, whereas serum specimens collected from patients on high-dose biotin therapy will contain biotin as well as biotin metabolites. The major metabolites of biotin are bisnorbiotin and biotin sulfoxide. With administration of 0.5, 2, and 20 mg biotin, more than 50% ingested dose is excreted as unchanged biotin.[22] Interestingly, biotin metabolites that preserve the ureido ring of biotin still bind streptavidin, but with reduced affinity. Therefore major biotin metabolites in human plasma (bisnorbiotin and biotin sulfoxide) should not cause a profound interference with biotin-based immunoassays as native biotin.[15]

Biotin Interferences in Thyroid Function Tests

Clinically significant interference of biotin in thyroid function tests is probably the most well-documented interference of biotin in clinical laboratory test results. Biotin shows positive interference with thyroxine (T_4), free thyroxine (FT_4), triiodothyronine (T_3), and free triiodothyronine (FT_3) assays because these are relatively small molecules (molecular weight of T_3 is 650.98 Da and molecular weight of T_4 is 776.87 Da) and immunoassays used for their analysis are competitive immunoassays. In contrast thyroid-stimulating hormone (TSH) is a large molecule (molecular weight, 28 kDa) that requires sandwich immunoassay format for analysis. As a result, biotin falsely lowers TSH value. However, the threshold serum biotin level at which clinically significant interferences are observed with thyroid function tests varies widely between each analyte as well as between each test manufactured by different diagnostic companies.

First Case Report of Biotin Interference

In 1996, Henry et al. published a case report of biotin interference in measurement of FT_4 and TSH using ES700 analyzer and immunoassays manufactured by Boehringer Mannheim, which uses streptavidin-coated tubes with binding capacity of 14–15 ng of biotin per tube. Cord blood measurement of a girl baby delivered vaginally at 39 weeks of gestation and weighing 3.92 kg showed a normal TSH and FT_4 levels but when the analysis was repeated 2 days later using a new blood sample, the TSH level was significantly reduced but FT_4 level was significantly elevated. Further investigation revealed that the baby was put on a vitamin formula (20 mg thiamine, 10 mg biotin, and 10 mg riboflavin) immediately after birth as a sibling has died of organic acidosis 2 years ago. However, no discrepancy in TSH and FT_4 values was observed between cord blood and blood collected from the infant on day 3 when a non–biotin-based immunoassay (Amerlite) was used. Further investigation using in vitro study by the authors showed biotin interference in the TSH at a level of

50 ng/mL, but for the FT_4 interference the threshold biotin level was 125 ng/mL. Moreover, biotin interference in the TSH assay was negative because it was a sandwich assay, whereas it was positive in the FT_4 measurement because it was a competitive assay. Biotin did not interfere in the cord blood measurement of TSH and FT_4 because the reference range of biotin in cord blood is 0.49–1.24 ng/mL. The package insert indicated that biotin did not interfere with TSH assay if present at a concentration less than 30 ng/mL. Moreover, the package insert also recommended that blood must be drawn 8 h after last biotin dose to avoid biotin interference in TSH assay. However, the authors commented that this recommendation may not be applicable to neonates because much longer waiting time may be needed to overcome biotin interference.[23]

Biotin Interference in Thyroid Function Tests: Other Case Reports

Case Report: A 3-year-old Chinese girl who had a history of recurrent generalized tonic-clonic seizure refractory to antiepileptic treatment after birth also suffered from several episodes of pneumonia, metabolic acidosis, and hypoglycemia. Further investigation indicated that the child may have propionic acidemia. Her condition stabilized after medical treatment but she still continued to have mild developmental delay and poor weight gain, which prompted the clinician to order thyroid function tests. By Roche Modular Analytics E170 and cobas e601 immunoassay analyzers (Roche diagnostics), her serum TSH level was 0.01–0.23 mIU/L (age-specific normal range, 0.9–6.5 mIU/L); FT_4 level, 12.6–17.0 pmol/L (normal, 12.0–22.0 pmol/L); and FT_3 level, 6.2 pmol/L (normal, 2.8–7.1 pmol/L). There was no known family history of thyroid disorder. Moreover, the child showed no symptom of hyperthyroidism. The clinician decided to reorder the thyroid profile tests but follow-up tests persistently showed low TSH but normal FT_4 concentrations. The chemical pathologist was concerned and initiated an investigation. The serial dilution using manufacturer-recommended diluent showed nonlinear recovery response and the TSH value approached 3.3 mIU/mL upon further dilution, which confirmed negative interference in serum TSH analysis.

When specimens were analyzed simultaneously using the Roche and Beckman Coulter analyzers, TSH concentration was 0.62 mIU/L by the Roche method but 3.96 mIU/L by the Beckman Coulter analyzer. However, FT_4 concentration was 75.9 pmol/L and FT_3 concentration was 14.0 pmol/L by the Beckman Coulter analyzer but significantly lower (FT_4 concentration, 15.5 pmol/L and FT_3 concentration, 4.5 pmol/L by Roche cobas e601 method). The FT_4 and FT_3 levels were found to be in euthyroid level using the cobas e601 analyzer, but the values obtained by using the Beckman Coulter analyzer were significantly higher than the upper limit of reference range. Further review of the patient's history revealed that upon diagnosis of propionic acidemia, she was put on special low-protein formula, oral biotin 10 mg, and levocarnitine 300 mg four times as daily supplementation. At that point, biotin interference in thyroid profile testing was considered.[24]

In the cobas e601 assay a biotinylated monoclonal TSH-specific antibody and a monoclonal TSH-specific antibody labeled with a ruthenium complex react to form a sandwich complex. After adding streptavidin-coated microparticles the complex becomes bound to the solid phase because of the interaction of biotin with streptavidin. The reaction mixture is aspirated into the measuring cell where the microparticles are magnetically captured onto the surface of the electrode. Unbound substances are then removed. Application of a voltage to the electrode then induces chemiluminescent emission that is measured by a photomultiplier. The intensity of the signal is roughly proportional to the TSH concentration in the sample. High biotin levels in serum (25 ng/mL or higher) can cause negative interference in the TSH assay by falsely lowering the signal by competing with biotinylated antibodies for the binding site on streptavidin. The in vitro study indicated that addition of exogenous biotin to achieve a concentration of 20 ng/mL caused negative interference in the TSH assay on the cobas e601 analyzer. However, the Beckman Access TSH assay does not use biotin-streptavidin interaction in the assay principle and is not subjected to biotin interference. The package insert for TSH assay manufactured by Roche Diagnostics indicates that for a patient receiving therapy with a high biotin dosage (over 5 mg per day), no specimen should be collected at least 8 h after the last biotin administration. However, the Beckman Access FT_4 and FT_3 competitive immunoassays utilize monoclonal T_4 antibody coupled to biotin and biotinylated T_3 analog, respectively, in their assay design. As a result, biotin causes positive interference (falsely elevated value) in both the FT_4 and FT_3 assays. Although Roche FT_4 and FT_3 assays also are biotin based, they were not affected significantly after ingestion of 10 mg of biotin. This case report clearly indicates that biotin interferences may vary from one manufacturer to another for the same analyte where biotin is used in the assay design.[24]

Wijerantne et al. also reported a case of a 1-week-old baby boy born with features of liver failure and lactic acidosis who was treated with 30 mg of biotin per day since day 2 of life. At 1 week, his thyroid function test results using Beckman DxI analyzer showed elevated FT_4 value >77.7 pmol/L (reference range, 25–70 pmol/L) and elevated FT_3 value of 24.9 pmol/L (reference range, 3.8–6.0 pmol/L) but a normal TSH level of 3.75 mU/L. This discordant assay results prompted further investigation. The analysis of the specimen using different methods (Abbott Architect analyzer and Centaur analyzer) showed normal thyroid profile. The authors suspected biotin interference in the measurement of FT_4 and FT_3 using the Beckman analyzer. To confirm the biotin interference, one of the authors ingested 30 mg of biotin and blood specimens were collected 0, 1, 2, 4, 8, and 25 h after ingestion of biotin. Serum specimens were tested on Beckman DxI for FT_4, FT_3, and thyroglobulin (Tg); the Abbott Architect analyzer for FT_4 and FT_3; and the Roche E170 analyzer for dehydroepiandrosterone sulfate (DHEAS), estradiol, testosterone, and ferritin. The authors reported falsely elevated FT_4 and FT_3 values using the Beckman DxI analyzer and falsely elevated testosterone, DHEAS, and estradiol values using the Roche E170 analyzer but negative interference in Beckman Tg assay and Roche ferritin assay. However, no interference was observed using the Abbott Architect analyzer (FT_4 and FT_3 assays) as biotin-streptavidin interaction is not used in the assay design. The authors observed maximum interference of biotin approximately 2 h after ingestion of biotin in all the analytes. The biotin interference lasted for approximately 5 h for DHEAS, estradiol, testosterone, and ferritin. However, biotin interference lasted for 25 h for FT_4, FT_3, and Tg using Beckman DxI analyzer. The magnitude of interferences also differed significantly among various analytes tested. The most dramatic interference was observed in the Roche estradiol assay (138-fold increase from the initial value).[25]

Case Report: A 32-year-old man with X-linked adrenomyeloneuropathy and chronic depression was treated with hydrocortisone 10 mg bid, fampridine 10 mg bid, paroxetine 20 mg once a day, and three capsules per day of an "investigational drug" showed suppressed TSH and elevated FT_4 and FT_3 values using Roche cobas e170 analyzer. Positivity of anti-TSH receptor antibody confirmed the diagnosis of Graves disease and the patient received an antithyroid medication (carbimazole 40 mg per day). After 5 weeks of therapy a slight improvement was noticed. Because his endocrinologist was on holiday, the patient was referred to the author's department. When thyroid function tests were repeated in the author's hospital laboratory using a different method (Siemens ADVIA Centaur), all thyroid function test results were within the normal reference range. The authors contacted the clinical trial center and discovered that the investigational drug was indeed biotin prescribed at a dose of 100 mg tid as a part of a trial of high-dose biotin in X-linked adrenomyeloneuropathy. The TSH value was 0.052 mIU/L as measured by the Roche analyzer that utilizes biotinylated antibody but the TSH value was 1.83 mIU/L as measured by the ADVIA Centaur analyzer that does not use biotinylated antibody in TSH assay design. Similarly FT_4 and FT_3 concentrations were 51.6 pmol/L and 9.7 pmol/L, respectively, using the Roche analyzer but 14.4 pmol/L and 5.6 pmol/L, respectively, using the ADVIA Centaur analyzer. The authors discontinued treatment with carbimazole immediately.[26] It is important to note that the TSH, FT_4, and FT_3 assays on the ADVIA Centaur analyzer marketed by Siemens do not use biotinylated antibody but the TSH, FT_4, and FT_3 assays on Dimension Vista analyzer also marketed by Siemens utilize biotinylated antibody and suffer from biotin interference if present in serum at high concentrations.

Case Report: A 60-year-old woman with a medical history of primary progressive multiple sclerosis was referred to the author's endocrinology department due to abnormal thyroid function test results (TSH, 0.02 mU/L; FT_4, >103 pmol/L; and FT_3, >46 pmol/L). The TSH receptor antibody levels were markedly elevated (>40 IU/L) (all tests measured using Siemens Dimension Vista), indicating Graves hyperthyroidism but the patient did not show any symptom of thyroid disorder. The thyroid function tests were performed by her neurologist because the patient experienced a short episode of diarrhea, which resolved spontaneously after a few days. Two months earlier, thyroid function tests performed by the same laboratory (using the Siemens Dimension Vista system) showed results within the reference range (TSH, 0.96 mU/L; FT_4, 12.7 pmol/L). The only change in the past 2 months was the initiation of treatment with high-dose biotin (100 mg tid). Because of inconsistency in her clinical presentation and thyroid function test results, identical blood samples were tested by two other laboratories. In one laboratory (using a Vitros system), thyroid function test results indicated the possibility of Graves disease (TSH, <0.02 mU/l; FT_4, 15.9 pmol/L; and FT_3, 4.7 pmol/L) but the other laboratory that used the Abbott Architect system showed normal thyroid profile (TSH, 1.66 mU/L; FT_4, 15.3 pmol/L; and FT_3, 4.7 pmol/L), consistent with lack of any symptom of hyperthyroidism in this woman. The authors suspected that abnormal thyroid function test results obtained using both the Vista analyzer (Siemens Diagnostics)

and Vitros analyzer (Ortho Clinical Diagnostics) were due to biotin interferences because both assays are biotin based, but thyroid function tests on Architect are not biotin based, hence no interference. The authors stopped biotin treatment for 2 days and repeated the thyroid function tests on the Siemens Dimension Vista system. This time all thyroid function test results were within the normal range (TSH, 2.24 mU/L; FT_4, 17.6 pmol/L), thus confirming that wrong diagnosis of Graves disease in the patient was due to interference of biotin in thyroid function test results.[27]

There are other reports of wrong diagnosis of Graves disease due to biotin interference in thyroid function tests. A 63-year-old Caucasian woman had a background of secondary progressive multiple sclerosis diagnosed 15 years earlier. Her neurologist later started her on biotin 100 mg three times daily and she showed abnormal thyroid function test results indicating Graves disease. Thyroid function tests performed 6 months earlier showed results within the reference interval. Biotin was stopped, and repeated thyroid function tests using the cobas analyzer, a Roche Diagnostics analyzer, and the Beckman Coulter DxI analyzer showed normal results 3 days later. The patient had noted symptomatic benefit of her multiple sclerosis after starting biotin, so biotin was reintroduced. The thyroid function test results after an overnight fast and 16 h after her last dose of biotin again showed evidence of biotin interference.[28]

Kummer et al. reported case reports of six children receiving high-dose biotin treatment in the context of inherited metabolic diseases who surprisingly showed laboratory results suggestive of Graves disease (elevated FT_4 and total T_3 levels but low levels of TSH using Roche assays). In addition, levels of antithyrotropin receptor antibodies were also elevated. Antithyroid medication was initiated in three children. However, only one child had symptoms attributable to hyperthyroidism (tachycardia, restlessness, and failure to thrive), but ultrasonographic scans of the thyroid, including Doppler flow studies, were unremarkable in all patients. The authors searched the literature and suspected abnormal thyroid function test results to be related to biotin interference. The authors discontinued biotin supplementation. The TSH and thyroid hormone levels were normalized 24–48 h after the discontinuation of biotin, whereas levels of antithyrotropin receptor antibodies took up to 7 days to normalize. The authors commented that high-dose biotin treatment can cause misleading laboratory results by fully mimicking the typical laboratory test results expected in patients with Graves disease.[29]

Biscolla et al. studied the effect of 10 mg of biotin supplement on thyroid function tests marketed by Roche Diagnostics. Blood was collected from 19 adult healthy volunteers before and after (3 and 24 h) oral ingestion of 10 mg of biotin. The mean TSH level was 2.84 mIU/L before taking biotin supplement but the mean value was falsely lowered to 1.66 mIU/L 3 hours after biotin ingestion. In contrast, FT_4 value was increased from a mean value of 0.8 ng/dL (before biotin ingestion) to 1.2 ng/dL 3 hours after ingestion of biotin. The mean level of total T_3 also increased from a mean baseline value of 116 ng/dL to 154 ng/dL. After 24 h, all values returned to baseline values.[30]

Case Report: A 55-year-old man with multiple sclerosis revealed markedly elevated thyroid hormone levels and suppressed thyrotropin (TSH). However, thyroid uptake and scan with 250 μCi of ^{123}I showed normal results. His medical history included progressive multiple sclerosis, psoriasis, hypertension, and spinal stenosis. He denied any symptoms of hyperthyroidism. Thyroid function tests were repeated showing significantly elevated levels of total T_3 (>650 ng/dL) and FT_4 (>7.8 ng/dL), but TSH level was suppressed (0.02 mIU/L) and the TSH binding-inhibiting antibody test was positive at a high concentration (36 IU/L). All these tests were performed using Roche Elecsys analyzer. These test results indicated diagnosis of Graves disease, which did not correlate with the clinical picture of the patient. Further testing showed low Tg level (3.9 ng/mL obtained by using the Beckman Access analyzer), with negative Tg antibodies, a finding inconsistent with Graves disease, and a bioassay for the thyroid-stimulating antibodies showed negative result. The sex hormone–binding globulin was in the normal range obtained by the Roche assay. When the patient was questioned further, he informed the clinician that he was taking 300 mg biotin every day (100 mg three times a day) for treatment of his multiple sclerosis. The patient was asked to stop the biotin treatment temporarily, and 2 weeks later, repeated thyroid function tests showed completely normal results (FT_4, 1.4 ng/dL; total T_3, 160 ng/dL; TSH, 0.78 mIU/L; TSH binding-inhibiting antibody, <1 IUL; and Tg, 21 ng/mL, using the same analyzer used for initial testing). The patient continued to feel well and biotin therapy was resumed. This case clearly shows misdiagnosis of Graves disease due to biotin therapy.[31]

Rulander et al. reported a rare incidence of interference of antistreptavidin antibody in biotin-based immunoassays in a 61-year-old man who was diagnosed with hyperthyroidism subsequent to multiple measurements of TSH and FT_4 on a Roche platform. He was treated with methimazole, but despite therapy, his TSH level was not elevated and the FT_4 level was

subnormal. An initial T-uptake result of less than 0.2 T_4 binding index and nonlinear TSH dilution suggested immunoassay interference. Treatment of specimen with heterophilic blocking tube reduced some but not all interference observed in the TSH measurement. When specimen was reanalyzed with Siemens Centaur analyzer, TSH value of 2.49 mIU/L was observed and when diluted the calculated TSH value correlated with the value obtained using undiluted specimen, indicating no interference in TSH assay. This value was reported to the clinician and unnecessary treatment for hyperthyroidism was discontinued. The TSH value was 1.3 mIU/L using the Vitros analyzer. However, when specimen was pretreated with streptavidin-linked agarose, the TSH values observed by the Roche and Vitros assays were 2.53 mIU/L and 2.18 mIU/L, respectively, indicating that interference could be eliminated by treating specimen with streptavidin agarose. The authors concluded that interference observed by the Roche and Vitros assays was due to the presence of endogenous antistreptavidin antibody in the patient's sera, which interferes with biotin-based TSH assays (Roche and Vitros assays).[32] Fortunately, the frequency of confirmed antistreptavidin antibody is only 0.6% (8/1375) in primary care setting and such antibody rarely causes assay interferences.[33]

Trambas et al. commented that biotin interference is not limited to thyroid function tests. Biotin interfered with thyroid function tests and other tests in a blood specimen obtained from a patient with multiple sclerosis who had received a single dose of 300 mg of biotin several hours before the blood sample was obtained. Biotin interferences caused falsely high results in competitive immunoassay (FT_4, FT_3, testosterone, estradiol, progesterone, DHEAS, and vitamin B_{12}), while biotin caused negative interference in sandwich immunoassays (TSH, prostate-specific antigen [PSA], parathyroid hormone [PTH], luteinizing hormone [LH], and follicle-stimulating hormone [FSH]). For example, using the biotin-based assay, vitamin B_{12} level was >1910 ng/L, whereas using a non–biotin-based assay, it was 518.4 ng/L. As expected, no interference was observed using non–biotin-based immunoassays. The authors used Roche assay that is biotin based and compared the results with those from non–biotin-based assays obtained by using the Abbott Architect analyzer manufactured by Abbott Laboratories.[34] Results are summarized in Table 4.4.

Piketty et al. reviewed the issue of false biochemical diagnosis of hyperthyroidism due to biotin interference in streptavidin-biotin–based immunoassays and commented that lack of clinical presentation with

TABLE 4.4
Effect of Biotin on Biotin-Based and Non–Biotin-Based Assays for Selected Analytes in Patients Taking 300 mg Biotin Supplement Where Blood was Collected Several Hours After Taking Biotin

Analyte*	Molecular Weight	Assay Format	Non-Biotin-Based Assay Architect/Abbott	Biotin-Based Assay Roche	Bias
FT_4	T_4: 776.87	Competitive	0.88 ng/dL 11.3 pmol/L	>7.8 ng/dL >100 pmol/L	Positive bias (>8.8-fold increase)
FT_3	T_3: 650.98	Competitive	292.2 pg/dL 4.5 pmol/L	1123.3 pg/dL 17.3 pmol/L	Positive bias (3.8-fold increase)
Testosterone	288.4	Competitive	291 ng/dL 10.1 nmol/L	1236 ng/dL 42.9 nmol/L	Positive bias (4.2-fold increase)
DHEAS	368.49	Competitive	244.4 µg/dL 6.6 µmol/L	>1000 µg/dL 27.1 µmol/L	Positive bias (>4.1-fold increase)
Estradiol	272.4	Competitive	19.8 pg/mL 73 pmol/L	200.5 pg/mL 40 pmol/L	Positive bias (10.1-fold increase)
Progesterone	314.4	Competitive	12.5 ng/mL 0.4 nmol/L	3943 ng/mL 125.4 nmol/L	Positive bias (315-fold increase)
Vitamin B_{12}	1335	Competitive	518.4 ng/L 380 pmol/L	>1910 ng/L >1400 pmol/L	Positive bias (>3.7-fold increase)
TSH	28 kDa	Sandwich	1.30 µIU/mL 1.30 mIU/L	0.02 µIU/mL 0.02 mIU/L	Negative bias (65-fold decrease)
PTH	12,500	Sandwich	35 pg/mL 2.8 pmol/L	7.5 pg/mL 0.6 pmol/L	Negative bias (4.7-fold decrease
PSA	28 kDa	Sandwich	0.6 ng/mL 0.6 µg/L	0.04 ng/mL 0.04 µg/L	Negative bias (4.7-fold decrease)
LH	29.5 kDa	Sandwich	1.4 U/L 1.4 IU/L	0.2 U/L 0.2 IU/L	Negative bias (7-fold decrease)
FSH	30 kDa	Sandwich	8.5 U/L 8.5 IU/L	0.04 U/L 0.4 IU/L	Negative bias (21.3-fold decrease)

*Molecular weight of each analyte is listed to show why some assays used were competitive immunoassays (lower molecular weight) while others were sandwich immunoassays (high molecular weight).
Values are given in both conventional units (first line) and SI unit (second line).
DHEAS, dehydroepiandrosterone sulfate; *FSH*, follicle-stimulating hormone; *FT₃*, free triiodothyronine; *FT₄*, free thyroxine; *LH*, luteinizing hormone; *PSA*, prostate-specific antigen; *PTH*, parathyroid hormone; *T₃*, triiodothyronine; *T₄*, thyroxine; *TSH*, thyroid-stimulating hormone.
Source of data: Trambas CM, Sikaris KA, Lu ZX, More on biotin treatment mimicking Graves' disease. *N Engl J Med*. 2016;375:1698.

laboratory test results, for example, test results indicating hyperthyroidism in a clinically euthyroid patient, should indicate the possibility of assay interference, including biotin interference.[35]

Biotin Falsely Lower Parathyroid Hormone Level

PTH maintains serum calcium homeostasis by directly altering calcium metabolism in bone and kidney and indirectly in the intestine. Testing of PTH is important because primary hyperparathyroidism is the leading cause of hypercalcemia in the outpatient setting and the

third most common of endocrine disorder. The PTH levels in a 60-year-old woman ranged from 90 to 336 pg/mL (reference range, 10–60 pg/mL), her calcium progressively increased from 10.3 to 11.5 mg/dL (reference range, 8.5–10.2 mg/dL) and a low 25-hydroxyvitamin D level of 27 ng/mL (reference range, 31–80 ng/mL) was also observed. Interestingly, when the patient was being worked up for surgery, her PTH level was undetectable, although her serum calcium level was still elevated. At that time, it was revealed that she was taking 1.5 mg biotin supplement daily for hair growth. After biotin was stopped the PTH test was repeated a month later and the PTH value was 197 pg/mL. To confirm the interference of biotin in PTH assay the patient was restarted on biotin and her PTH level was again undetectable. The Roche PTH assay utilizes biotinylated antibody and is susceptible to biotin interference. The authors also described a second case report where PTH levels were undetectable when the patient was taking 5 mg of biotin supplement daily for neuropathic pain.[36]

Case report: A 64-year-old female with long-standing end-stage renal disease was evaluated for management of renal osteodystrophy with particular concern for adynamic bone disease due to low intact PTH level, intermittently increased serum calcium level, and severe osteoporosis. However, her mildly increased serum alkaline phosphatase activities were inconsistent with low bone turnover. When PTH was analyzed simultaneously using Roche Elecsys analyzer at the author's laboratory and at a reference laboratory using Siemens Immulite 2000 analyzer, discrepant values (48 ng/L using Elecsys and 786 ng/L using Immulite 2000) were observed. The authors commented that increased intact PTH level of 786 ng/mL from the reference laboratory was more consistent with the clinical picture of the patient, indicating that the value obtained in the author's laboratory was falsely reduced because of assay interference. Further investigation revealed that the patient was taking 10 mg biotin per day for the past 2 years for restless leg syndrome. An aliquot of the specimen was sent to another reference laboratory for determination of biotin concentration and the value was 4.8 ng/mL, approximately 10 times higher than the normal serum biotin level (0.2–0.5 ng/mL). The Roche package insert indicted that intact PTH tests are unaffected when serum biotin concentration is less than 50 ng/mL (less than 10% difference of the target value). However, the bioassay used for measuring biotin underestimated other biotinylated molecules present in the serum. The author concluded that Elecsys intact PTH assay utilizes biotin in the assay design and is affected by high serum biotin level but the Immulite 2000 assay is not biotin based and hence no interference. The authors commented that biotin interference in intact PTH measurement in the serum using the Roche Elecsys assay persisted even when the blood specimen was collected from the patient more than 8 h after biotin administration.[37]

Ranaivosoa et al. reported significant interference of biotin in TSH, FT_4, and PTH testing and commented that kidney failure combined with therapeutic biotin may cause seriously misleading results in immunoassays using biotin-streptavidin mechanism.[38]

Biotin Interference in Troponin Assays

Biotin falsely lowers troponin T levels in tests using the Roche analyzer. For example, troponin T value was reduced from 39 ng/L to 3 ng/L in the presence of 500 ng/mL biotin. The high vulnerability of the troponin T assay to biotin interference is clinically concerning because high-dose biotin therapy can lower troponin T results to below the 99th percentile cutoff, and in the early investigation of a potential myocardial infarction, this might lead to inappropriate discharge of a patient and potentially missing a diagnosis of myocardial infarction. Therefore a false-negative troponin I level in the context of atypical chest pain and nonspecific ECG change may have dire clinical consequences because values may be falsely lower even with a biotin concentration of 31.3 ng/mL.[15] The FDA reported death of a person due to interference of biotin in troponin immunoassay.[8]

Several troponin I assays available commercially are biotin based. Therefore negative interference in troponin I value is expected if biotin is present in the serum at elevated concentrations. Willeman et al.[39] reported negative interference of biotin at a concentration of 300 ng/mL in the troponin I assay using the Vista analyzer (Siemens Healthineers). However, troponin I assay in the Vitros analyzer did not show any significant interference of biotin at a serum concentration of up to 200 ng/mL.[40] Collinson et al.[41] evaluated clinical performance characteristics of the highly sensitive Abbott cardiac troponin I assay and concluded that the analytical performance of the new assay satisfies the criteria for a high-sensitivity troponin I assay. Because the Abbott cardiac troponin assay is a non–biotin-based assay, it is not affected by high biotin concentrations in serum.

Biotin Interferences in Multiple Immunoassays

Li et al. investigated the effect of biotin on the performance of hormone and nonhormone assays using six healthy adults (two women and four men; mean age, 38 years) who ingested 10 mg biotin (>300 times higher than the daily recommended allowance) for 7 days. This particular dose was selected because 10 mg biotin supplement is available over the counter and a biotin dose of 10 mg and higher is also prescribed in dermatologic, medical, and neuropsychiatric conditions. The authors analyzed 11 analytes, including 9 hormones, by collecting blood before taking biotin supplement, after a week of taking biotin supplement (day 7, 2h after the last dose of biotin), and 1 week after discontinuation of biotin supplement (day 14; washout period of 1 week). The mean baseline serum biotin concentration was 774 pg/mL (0.77 ng/mL) before taking biotin

TABLE 4.5				
Effect of 10 mg Biotin Supplements on Biotin-Based and Non-Biotin-Based Immunoassays				
Diagnostic Company	Analyzer Used	Biotin-Based Assay?	Individual Assay	Effect
Abbott Laboratories	Architect	No	FT_4, total T_3, total T_4, PSA, PTH, TSH, ferritin, prolactin	No effect
Abbott Laboratories	Architect	Yes	25-Hydroxyvitamin D	No effect
Roche Diagnostics	cobas e602	Yes	FT_4, total T_3, FT_3	Falsely high values (competitive immunoassays)
Roche Diagnostics	cobas e602	Yes	Total T_4	No effect
Roche Diagnostics	cobas e602	Yes	25-Hydroxyvitamin D	Falsely high value (competitive immunoassay)
Roche Diagnostics	cobas e602	Yes	TSH	Falsely low value (sandwich immunoassay)
Roche Diagnostics	cobas e602	Yes	PTH, ferritin, NT-proBNP, prolactin	No effect (sandwich immunoassay)
Ortho Clinical Diagnosis	Vitros 5600	No	FT_4, total T_3, total T_4	No effect
Ortho Clinical Diagnosis	Vitros 5600	Yes	PTH, TSH, NT-proBNP	Falsely low values (sandwich immunoassay)
Ortho Clinical Diagnosis	Vitros 5600	Yes	Ferritin	No effect (sandwich immunoassay)
Siemens Healthineers	Vista	Yes	FT_3	Falsely high value (competitive immunoassay)
Siemens Healthineers	Vista	Yes	FT_4, ferritin	No effect (competitive immunoassay)
Siemens Healthineers	Vista	No	Total T_4	No effect
Siemens Healthineers	Vista	Yes	PSA, ferritin, prolactin, NT-proBNP	No effect (sandwich immunoassay)
Siemens Healthineers	Centaur	No	Total T_3	No effect
Siemens Healthineers	Centaur	Yes	PTH	No effect
Siemens Healthineers	Immulite	No	Prolactin	No effect

FT_3, free triiodothyronine; FT_4, free thyroxine; *NT-proBNP*, N-terminal prohormone brain natriuretic peptide; *PSA*, prostate-specific antigen; *PTH*, parathyroid hormone; T_3, triiodothyronine; T_4, thyroxine; *TSH*, thyroid-stimulating hormone.

Li D, Radulescu A, Sherstha RT, Root M, et al. Association of biotin ingestion with performance of hormone and non-hormone assays in healthy adults. *J Am Med Assoc*. 2017;318:1150–1160.

supplement. However, serum biotin concentration was increased to more than 3600 pg/mL in all subjects at day 7 of taking biotin supplement but then reduced to baseline value (mean value, 1090 pg/mL) after 14 days (7 days after discontinuing biotin supplement). The analytes tested by the authors include TSH, total T_4, total T_3, FT_4, FT_3, PTH, prolactin, N-terminal prohormone brain natriuretic peptide (NT-proBNP), 25-hydroxyvitamin D, PSA, and ferritin by using various assays (23 biotinylated assays and 14 nonbiotinylated assays) and analyzers (Vitros, Centaur, Vista, cobas e602, and Architect) obtained from different diagnostic companies (Ortho Clinical Diagnosis, Siemens, Roche Diagnostics, and Abbott Laboratories). Overall the authors observed that 5 out of 8 biotinylated competitive immunoassays (63%) showed falsely elevated results, while only 4 out of 15 biotinylated sandwich assays (27%) showed falsely lower values. As expected, all 14 nonbiotinylated immunoassays showed no biotin interference. These results are presented in Table 4.5.

Magnitude of biotin interference for the same analyte also varied significantly among different biotinylated immunoassays obtained from different manufacturers. For example, biotin ingestion (10 mg per day for 7 days) was associated with falsely decreased TSH level (mean decrease, 0.72 mIU/L, 37% reduction from baseline) measured using Roche cobas e602 analyzer, although all results remained within the euthyroid reference range. The interference in TSH measurement was more significant in the Vitros 5600 analyzer (Ortho Clinical Diagnosis) where TSH value decreased by a mean of 1.67 mIU/L (94% decrease) and all results were falsely decreased to below the reference range. Biotin ingestion was also associated with statistically significant false increases (positive interference) in the Roche total T_3, FT_4, and FT_3 assays performed on the cobas e602 analyzer and also FT_3 assay using the Dimension Vista 1500 analyzer (Siemens Healthineers). In the Roche cobas e602 assay the total T_3 concentrations were falsely increased by a mean of 0.85 ng/mL, whereas in the Siemens Vista assay the total T_3 concentrations were falsely increased by a mean of 0.78 ng/mL. Biotin ingestion also significantly decreased the PTH values when measured by the Vitros PTH assay (mean decrease, 25.8 pg/mL, 61% reduction from baseline value).

Biotin ingestion was associated with falsely lower NT-proBNP values (using Vitros 5600 analyzer) in all participants. However, the NT-proBNP assay performed using cobas e602 and Vista analyzers (using biotin-based immunoassays) was not affected when volunteers ingested 10 mg biotin supplement. However, biotin ingestion falsely increased 25-hydroxyvitamin D assay using cobas e602 analyzer but the Architect assay also using biotinylated antibody showed no interference. Biotin ingestion has no effect on prolactin measured using biotin-based immunoassays on the cobas e602 and Vista analyzers. The authors commented that differential biotin interference tolerance among assays is likely due to the amounts of biotinylated antibodies or analogs used, the availability of streptavidin-binding sites, and other factors in the assay reagents (e.g., streptavidin-coated magnetic particles).[42]

Ali et al. evaluated the effect of biotin on various analytes analyzed by the Vitros 5600 analyzer (Ortho Clinical Diagnostics) using in vitro supplementation of biotin and reported that biotin (up to 200 ng/mL) did not significantly affect troponin I and hepatitis A virus antibody assays. Moreover, at a relatively low concentration (up to 12.5 ng/mL) biotin resulted in <10% bias in creatinine kinase MB, β-human chorionic gonadotropin (hCG), α-fetoprotein, cortisol, and ferritin assays but at high biotin concentrations (200 ng/mL), these assays were also affected. However, TSH and prolactin were susceptible to significant interferences from biotin at a relatively lower concentration. Biotin >6.25 ng/mL significantly affected TSH (>20% bias) assay. Prolactin was significantly affected even at low levels of biotin (1.5 ng/mL). The authors concluded that the most significantly affected assays were TSH and prolactin.[40]

Willeman et al. investigated biotin interferences in 16 routine immunoassays using luminescent oxygen channeling assay (LOCI) on Siemens Dimension Vista analyzer. The authors observed no interference for biotin concentrations between 50 and 200 ng/mL. However, interferences were observed at 300 ng/mL for troponin I (negative interference) and FT_3 (positive interference). Above 400 ng/mL, significant interferences were observed with digoxin (positive interference), NT-proBNP (negative interference), TSH (negative interference), and FT_3, FT_4, progesterone, and estradiol (all positive interferences). However, no interference was observed in β-hCG, LH, ferritin, and myoglobin, as well as free and total PSA assays. As expected, the authors observed positive interferences with competitive immunoassays but negative interferences using sandwich immunoassays.[39] In our experience, when drug aliquots of drug-free serum pool were supplemented with various amounts of biotin, apparent digoxin concentration was observed at a biotin concentration of 50 ng/mL and higher. The bulletin issued by Siemens also stated that the threshold of biotin interference in the LOCI digoxin assay is 50 ng/mL. Interestingly, when aliquots of the digoxin pool were further supplemented with biotin, a statistically increased digoxin level was observed only with 250 ng/mL biotin, indicating that it is unlikely for a person taking 5 or 10 mg of biotin supplement to show any significant effect in therapeutic drug monitoring of digoxin using the LOCI digoxin assay.[43]

Piketty et al. also studied the effects of biotin on various assays: FT_3, FT_4, TSH, PTH, cortisol, FSH, LH, prolactin, C-peptide, and 25-hydroxyvitamin D in 20 patients receiving high-dose biotin; 8 healthy volunteers taking a single dose of 100, 200, or 300 mg biotin; and 3 authors taking 15 or 30 mg biotin per day. The authors observed interference of biotin (>10% change over baseline) at a serum biotin level of 30 ng/mL or more using assays marketed by Roche Diagnostics. However, susceptibility of assays to biotin interference varied widely. Clinically misleading results were observed for 25-hydroxyvitamin D and PTH at a serum biotin concentration of 169 ng/mL and higher. The threshold values for biotin interferences in various

TABLE 4.6
List of Diagnostic Companies Marketing Biotin-Based and Non-Biotin-Based Immunoassays

Diagnostic Company	Platform	Biotin-Based Assays	High Risk for Biotin Interference	Comments
Roche	Elecsys/ cobas/ Modular	81	44	Results may be affected in individuals taking 5 mg biotin or more per day. At 31.3 ng/mL biotin level, most assays are unaffected, except for TSH, troponin T, anti-TPO, and anti-TSHR. However, at 500 ng/mL biotin level, all assays are affected.
Ortho Clinical Diagnosis	Vitros	30	28	10 mg biotin supplement per day affected LH, FSH, and TSH values. After discontinuation of biotin supplement, results returned to normal in 2 days. Taking 10 mg supplement per day also lowered PTH result in one patient. Hepatitis B e antigen and antibody tests were also affected at high biotin levels.
Siemens	Dimension	21	6	Usually tests are affected at biotin levels of 200–400 ng/mL but after taking 10 mg biotin supplement, only FT_3 assay was affected.
Siemens	Immulite	60	6	No effect in prolactin assay after ingestion of 10 mg biotin.
Siemens	ADVIA Centaur	23	7	Older PTH and troponin I assays were affected but the reformulated PTH and troponin I ultra assays do not show significant interference.
Beckman Coulter	Access/DxI	15	6	Six immunoassays (FT_3, FT_4, total T_3, thyroglobulin, CA 125, and CA 15-3) are at high risk for biotin interferences
Abbott	Architect	4	0	Not applicable
DiaSorin	Liaison XL	0	0	Not applicable

FSH, follicle-stimulating hormone; *FT_3*, free triiodothyronine; *FT_4*, free thyroxine; *LH*, luteinizing hormone; *PTH*, parathyroid hormone; *T_3*, triiodothyronine; *T_4*, thyroxine; *TPO*, thyroid peroxidase; *TSH*, thyroid-stimulating hormone; *TSHR*, thyrotropin hormone receptor.

biotin-based immunoassays are listed in Table 4.6. The Roche prolactin assay was the least affected.[19]

Biotin can also interfere with urine-based testing because approximately 50% biotin is secreted in urine unchanged. In general, 20% of ingested biotin dose is excreted in urine within 4 h. Therefore ingestion of high doses of biotin may interfere with urine-based assays. In one study the authors investigated potential interference of biotin in qualitative point of testing for hCG both in vitro and in vivo and showed that only the QuickVue point of care device for detecting hCG was producing invalid result because the control line was very faint in the presence of significant amount of biotin but other devices (Alere 20, Alere 25, Icon 20, Osom, QuPID, and SureVue) were not affected. When volunteers ingested 10 mg biotin supplement each day for 7 days, invalid hCG test result using QuickVue was observed in the urine of some subjects because of a very faint control line.[44]

Diagnostic Companies Using Biotin-Based Assays: A Review

Biotin interference is only applicable to biotin-based immunoassays manufactured by several diagnostic companies.[45] Theoretically all biotin-based immunoassays are vulnerable to biotin interferences. However, the threshold biotin serum concentration varies widely for causing clinically significant interferences. Homes et al. reviewed the current manufacturers' instructions for use of 374 immunoassays performed by eight of the most popular immunoassay analyzers in the United States and commented that 221 of these assays are biotin-based assays. Out of 221 assays, 82 have manufacturer-reported interference threshold for biotin interferences (greater than 10% change in test result) at a concentration less than 51 ng/mL. The authors also identified 44 high-risk assays performed on the Roche Elecsys system.[46] The number of high-risk assays susceptible to biotin interferences is listed in Table 4.6.

Because Roche has utilized biotin in almost all its assays that are generally used in clinical laboratories, major attention has been focused on the effect of biotin on various Roche assays. Picketty et al. evaluated the impact of biotin on various assays performed using cobas e411 and observed that most assays were affected in subjects taking 100 mg biotin supplement and the magnitude of interference was directly proportional to serum biotin concentrations. However, a lower dosage of biotin (15 mg/day) showed 15%–20% falsely elevated 25-hydroxyvitamin D value. A 30-mg dose resulted in significant interferences in DHEAS, testosterone, and ferritin assays. The prolactin assay was the least affected.[19]

Trambas et al. published a detailed report on their findings on biotin interferences in Roche immunoassays and observed that both competitive and sandwich assays were minimally affected (<5% analytical bias) in the presence of 15.6 ng/mL biotin, with the exception of TSH and troponin T that demonstrated almost 10% negative bias at that biotin concentration. Moreover, anti-Tg, anti–thyroid peroxidase (TPO), and anti–thyrotropin receptor (TSHR) showed greater than 10% positive bias at 15.6 ng/mL biotin concentration in serum. Again at 31.3 ng/mL biotin concentration in serum, most Roche assays showed less than 5% analytical bias, except for anti-TPO and anti-TSHR (showed twofold positive bias), anti-Tg (50% positive bias), TSH (approximately 25% negative bias), and troponin T (20% negative bias). Some assays such as FT_4, FT_3, LH, and FSH showed statistically significant changes from baseline values, but analytical bias was less than 10%. The authors commented that serum biotin level of 15.6 and 31.3 ng/mL is stimulation of 5–10 mg biotin intake. Therefore most Roche assays should not be significantly affected after ingestion of 5–10 mg of biotin. However, at biotin concentrations of 500 ng/mL and higher, all Roche assays are affected. Of the sandwich assays, TSH, insulin, prohormone brain natriuretic peptide (proBNP), troponin T, and free PSA were grossly affected at 500 ng/mL serum biotin levels. In addition, FSH, PSA, PTH, and adrenocorticotropin hormone assays were also severely affected (more than 90% reduction in analyte concentrations). In contrast, CA 19-9 (approximately 40% reduction) and CA 15-3 (approximately 25% reduction) were less affected at 500 ng/mL serum biotin level. For competitive immunoassays (at 500 ng/mL serum biotin concentration), digoxin (>15-fold increase), anti-TPO, anti-Tg (up to 50-fold increase), and anti-TSHR assays showed extreme susceptibility to interference. Other competitive assays were also significantly affected at 500 ng/mL serum biotin concentration; for example, FT_4 and

testosterone showed fivefold and sevenfold increases, respectively, and FT_3, estradiol, and DHEAS showed threefold to fourfold increases. In summary, among all the Roche immunoassays, TSH and troponin I assays are highly susceptible to biotin interference but other assays such as CA 125, CA 19-9, CA 15-3, carcinoembryonic antigen, and hCG are relatively resistant to biotin interference. Competitive immunoassays are subject to more marked interference when concentration of the analyte is very low. In contrast, sandwich immunoassays are more consistently affected across the reference range.[15] Samarasinghe et al.[6] commented that all 81 assays on the Roche Elecsys analyzer are biotin based and theoretically vulnerable to biotin interferences.

Roche Diagnostics has completed a pharmacokinetic study where 54 subjects took various amounts of biotin and commented in its bulletin that consuming 30–60 μg of biotin, which represents the typical amount present in multivitamin formulations, has no effect on Roche assays. However, some people may take 5–10 mg biotin as supplement for healthy hair and nail growth. The study shows that after taking 5 mg supplement, 100% subjects showed blood biotin level below 30 ng/mL after 3.5 h of taking supplement. Similarly 8 h after taking 10 mg supplement, serum biotin levels were below 30 ng/mL in 100% subjects. Therefore the potential risk of biotin interferences only occurs with megadoses of biotin (5000 μg/5 mg or higher) and Roche recommends that specimens must be collected at least 8 h after the last dose if a person is taking 5 mg biotin supplement. Based on the Nielsen FDM data, Roche Diagnostics commented that although sale of biotin has been increased by 6.4% from July 2015 to June 2016, it has slowed down by 3.3% from July 2016 to June 2017. Moreover, lower doses (2.5 mg or lower) make up the majority of biotin sales and such doses have no effect on biotin-based assays.[47] There is some evidence that biotin supplement at a dosage of 2.5 mg per day may have some beneficial effect on nail growth, but there is no evidence that higher biotin doses have any additional benefits.[48]

In an article based on interviewing experts in the field of laboratory medicine, the author reported that of about 628 million thyroid tests supplied by Roche Diagnostics in a year, the company received 14 biotin-related case reports. However, James Freeman, senior director of immunoassays development for Siemens, commented in the same article that there is no dispute that biotin when taken in megadoses as supplement can cause interferences in immunoassays that use biotin-streptavidin architecture. Dr. Andre Valcour, director of the Esoteric Immunoassay, Allergy, Coagulation, Toxicology and Biological Monitoring departments at

LabCorp's Center for Esoteric Testing, Burlington, NC, commented that he has been surprised at how many participants, based on the belief that biotin's role as a key contributor to keratin will help improve hair and nails, take biotin supplements and they did not know anything about the potential effect of biotin megadoses on laboratory testings.[49]

Some of the assays on the Vitros analyzer marketed by Ortho Clinical Diagnostics also utilize biotin in the assay design. Ali et al.[40] evaluated the effect of biotin on various analytes analyzed by the Vitros 5600 analyzer and observed that TSH and prolactin assays are most vulnerable to biotin interference. Batista et al. reported that when a patient ingested 10 mg biotin supplement per day, the LH, FSH, and TSH values were falsely decreased, whereas values of estradiol, folate, and progesterone were falsely increased using Vitros analyzer. After discontinuation of biotin supplement, results returned to normal in 2 days.[50] Aguirre reported significant interference of biotin in hepatitis B e antigen and hepatitis B e antibody assay for application on Vitros analyzer.[51]

Some of the assays marketed by Siemens Healthineers also use biotin in the assay design but most assays only show significant interferences at much higher biotin concentrations (200–400 ng/mL).[6] After taking 10 mg biotin supplement, only FT_3 assay was affected.[42] According to Siemens' literature, digoxin assay is the most sensitive to biotin interference, as the noninterfering threshold is 50 ng/mL; however, in our experience, falsely elevated digoxin values were observed at 250 ng/mL after in vitro supplement of biotin in an aliquot of serum digoxin pool.[43] The test least affected is TSH, as the threshold for noninterference is 500 ng/mL or lower biotin levels in serum. Other affected assays (troponin I, estradiol, ferritin, and prolactin) may show significant interference at a serum biotin concentration exceeding 100 ng/mL.

Beckman Coulter issued a medical device safety alert in May 2017 that identified four immunoassays that are susceptible to biotin interference at a biotin concentration of 100 ng/mL or higher. However, Holmes et al. commented that six immunoassays (FT_3, FT_4, total T_3, Tg, CA 125, and CA 15-3) manufactured by Beckman Coulter (Access/DxI analyzer) are at high risk for biotin interferences.[47] Lim et al.[52] reported that biotin interferes with thyroid hormone and Tg tests but not TSH measurement using the Beckman Access analyzer. Correlation between ingested biotin doses and affected biotin-based immunoassays is summarized in Table 4.7. Moreover, the threshold of serum biotin concentration that affects assays manufactured by different diagnostic companies also vary significantly. In Table 4.8 the threshold biotin concentrations required to cause significant interference in most susceptible tests manufactured by different diagnostic companies are given.

TABLE 4.7
Some Examples of Correlation Between Biotin Dose (1.5–30 mg) and Affected Immunoassays Manufactured by Different Diagnostic Companies Based on In Vivo Studies

Biotin Dosage per Day (mg)	Population	Roche Assays	Ortho Assays	Siemens Assays	Beckman Assays	References
10	Adult (healthy volunteers)	FT_4, FT_3, total T_3, TSH, 25-hydroxyvitamin D	TSH, PTH, NT-ProBNP	FT_3	No data	42
10	3-Year-old girl	FT_4, FT_3, TSH	No data	No data	No data	24
10	64-Year-old female	Intact PTH	No data	No interference in Immulite 2000 assay	No data	37
30	1-Week-old boy	No data	No data	No data	FT_4, FT_3	25
30	Adult (healthy volunteer)	DHEAS, testosterone, estradiol, ferritin	No data	No data	FT_4, FT_3	25
1.5	Adult female	PTH	No data	No data	No data	36
5	Adult patient	PTH	No data	No data	No data	36

DHEAS, dehydroepiandrosterone sulfate; *FSH*, follicle-stimulating hormone; *FT_3*, free triiodothyronine; *FT_4*, free thyroxine; *NT-ProBNP*, N-terminal prohormone brain natriuretic peptide; *PTH*, parathyroid hormone; *T_3*, triiodothyronine; *TSH*, thyroid-stimulating hormone.

TABLE 4.8
Threshold Biotin Concentration for Biotin Interference in Assays Manufactured by Different Companies

Diagnostic Company	Biotin Threshold Value (ng/mL)	Comments
Roche	30	The assays affected at >30 ng/mL are anti-TPO and anti-TSHR (twofold positive bias), anti-Tg (50% positive bias), TSH (approximately 25% negative bias), and troponin T (20% negative bias). However, at 500 ng/mL and higher biotin concentration, all Roche assays are affected.
Siemens	50	According to Siemens, digoxin tests may be affected at biotin concentrations >50 ng/mL, but in our experience, significant interference was observed at 250 ng/mL. The test list affected is TSH, as the threshold is 500 ng/mL. Other assays (troponin I, estradiol, ferritin, and prolactin) may show significant interference at a serum biotin concentration exceeding 100 ng/mL.
Beckman	100	FT_3, FT_4, total T_3, Tg, CA 125, and CA 15-3 assays are most susceptible to interference.
Ortho	>6.25	According to an in vitro study, TSH values were affected at serum biotin levels exceeding 6.25 ng/mL. However, other assays were not affected unless biotin level exceeds 200 ng/mL.

FT_3, free triiodothyronine; FT_4, free thyroxine; T_3, triiodothyronine; Tg, thyroglobulin; TPO, thyroid peroxidase; TSH, thyroid-stimulating hormone; $TSHR$, thyrotropin hormone receptor.

APPROACHES TO ELIMINATE BIOTIN INTERFERENCE

The possibility of interference in an assay may be suspected from any observation or a combination of various observations.[53]

- Lack of correlation between test results and clinical presentation of the patient is a strong suspicion of assay interference because clinicians must treat patients but not laboratory test results. For example, observation of low TSH and elevated FT_3 and FT_4 levels indicating hyperthyroidism in a euthyroid patient is an indication of assay interference, including biotin interference.
- Comparison of physiologic dependent variables: Lack of the usual balance between the hormones and its regulating factor, for example, indication of a very rare syndrome of inappropriate secretion of TSH based on laboratory test results in a euthyroid patient.
- Generation of implausible laboratory test results: Extreme deviation from values expected in a pathologic condition is an indication of assay interference.
- Nonlinearity after dilution is an indication of assay interference, including biotin interference.
- Discordant results by different assays for the same analyte are an indication of assay interference.

One approach to eliminate biotin interference in clinical laboratory tests is to use an alternative assay platform that does not use biotin in the assay design. Currently, most assays marketed by Abbott Laboratories do not use biotin in the assay design and only four assays that are biotin based are not affected by high levels of biotin. DiaSorin Inc also has no assay that is biotin based.[46] Various approaches to overcome biotin interferences in immunoassays are summarized in Table 4.9.

However, for laboratories using assay platform that has biotin-based immunoassays, an alternative approach should be considered. The major problem is that there is no easily available assay to measure biotin concentration in serum. Therefore specimens containing high biotin concentrations cannot be screened to identify problem specimens. Unless a patient discloses his or her use of megadose of biotin, there is no way for the clinician or laboratory personnel to know that test results using biotin-based assays may be inaccurate. Several investigators have explored various methods to eliminate biotin interference. One simple approach is serial dilution of the specimen where nonlinearity upon dilution is an indication of the potential presence of interfering substance in the specimen. This approach, which is commonly used in clinical laboratories to eliminate hook effect and interferences of heterophilic antibodies in sandwich type immunoassays, has also been used by some investigators to overcome biotin interference. For example, Meany et al. confirmed negative interference of biotin in the intact PTH assay using

TABLE 4.9	
Approaches to Overcome Biotin Interference in Immunoassays	
Approach to Overcome Biotin Interference	**Comments**
Use immunoassays that do not use biotin in the assay design	This is probably the best approach. Currently all assays marketed by DiaSorin do not use biotin/streptavidin interaction. Moreover, only four assays on Abbott's Architect platform use biotin but they are not subjected to biotin interference.
Serial dilution	This is the simplest approach for laboratories using biotin-based immunoassays. Nonlinearity during dilution is an indication of biotin interference. However, serial dilution may not completely eliminate biotin interference.
Waiting for biotin clearance	Biotin has a short half-life and after taking 5 mg biotin supplement, waiting for 8 h may be sufficient to eliminate biotin interference. The American Thyroid Association recommends waiting for at least 2 days after cessation of high-dose biotin therapy before drawing blood for thyroid testing. After taking 100–300 mg biotin, waiting for at least 3 days is advised, but for individuals with renal failure, biotin, which is cleared renally, may cause interference for up to 15 days after the last dose. However, for acutely ill patients, no such waiting period is possible.
Remove biotin by pretreatment of specimen with streptavidin-coated microparticles	Incubation of serum with streptavidin-coated microparticles depletes up to 1000 ng/mL of biotin from the specimen, and such pretreatment eliminates biotin interference in both competitive and sandwich immunoassays that utilize biotin in the assay design.

the Roche Elecsys analyzer by dilution study. In the undiluted specimen, intact PTH value was 48 ng/L but after 1:20 dilution the intact PTH value calculated was 567 ng/L.[37] In another report, when specimens were analyzed simultaneously using Roche and Beckman Coulter analyzers, TSH concentration was 0.62 mIU/L by the Roche method but 3.96 mIU/L by the Beckman Coulter analyzer, indicating biotin interference in Roche method, but serial dilution with a validated assay diluent showed nonlinear recovery response and the TSH concentration approached 3.3 mIU/L upon further dilution using the Roche method.[24] The features of nonliner dilution include the following:

- Higher concentration of the analyte in the diluted specimen than that in the original specimen when the analyte is measured using sandwich immunoassay where biotin shows negative interference.
- Lower than expected concentration of the analyte in the diluted specimen compared with that in the original specimen when the analyte is measured using competitive immunoassay where biotin shows negative interference.

However, it is important to note that this nonlinearity during dilution is not only observed due to biotin interference but also a general feature of any interference.

Another approach to overcome biotin interference in biotin-based assays is to wait sufficient time for biotin to be eliminated from the body. Roche Diagnostics recommend waiting for 8 h before blood collection in patients taking 5 mg of biotin supplement. The American Thyroid Association Guidelines recommend discontinuation of biotin at least 2 days before conducting thyroid function tests.[54] Maximum interference of biotin is observed 2 h after biotin ingestion and biotin has a relatively short half-life of 1.8 h after ingesting a modest dose. Based on the pharmacokinetic data, waiting for 8 h after 10 mg/day biotin supplement should be sufficient to eliminate biotin interference.[17] Sulaiman reported significant interference of biotin in thyroid function tests using Roche assays in three adult patients taking high-dose biotin (1–10 mg/kg body weight daily) to treat their inherited metabolic disorder. However, thyroid function tests using Abbott Architect i 2000 analyzer showed normal results. When the author repeated thyroid function tests using Roche assay 13 h after taking biotin, all values were normal.[55] However, waiting for at least 3 days before undergoing laboratory tests is recommended for patients taking pharmaceutical dose of biotin (100–300 mg).[50,56] It may be reasonable to wait for a week before testing in pediatric patients taking between 2 and 15 mg/kg biotin daily.[23,29,57] Biotin is primarily excreted in urine and thus moderately impaired renal function is likely to reduce clearance of biotin and duration of potential biotin interference in biotin-based immunoassays. For example, PTH level underestimation has been reported up to 15 days in patients with end-stage renal disease taking 10 mg biotin.[19,37]

One of the major problems of waiting for enough time so that biotin is excreted in sufficient amount from the body and does not interfere with assays is that such waiting period is not possible in acutely ill patients. For example, biotin interferes with troponin T and biotin-based troponin I assays, but it is not possible to delay blood test in a patient admitted to the emergency department with a suspected myocardial infarction.

Currently, the best approach to overcome biotin interference in biotin-based immunoassays is pretreatment of the specimen with streptavidin-coated microparticles to deplete biotin present in specimens. In one study the authors used a streptavidin-coated magnetic microparticle suspension (0.72 mg/mL) obtained initially from unused Roche Elecsys immunoassays and later from used reagent packs when defined by analyzer as "empty." The authors prepared a pool of microparticles and stored it at 4°C. The authors successfully removed 250 ng biotin from a 250-μL specimen (biotin concentration in serum 1000 ng/mL) by incubating serum with immobilized magnetic microparticles coated with streptavidin (V-bottom tubes containing microparticle suspension were inserted in V-bottom plate magnet to immobilize magnetic particles and then excess fluid in suspension was removed by manual pipetting before adding specimen). Such treatment was capable of depleting biotin from specimens and as a result biotin interference was eliminated in both competitive and sandwich immunoassays, which incorporated biotin in assay design even in patients with multiple sclerosis taking 300 mg biotin daily.[58]

Piketty et al. also described a similar approach to overcome biotin interference in biotin-based immunoassays. The authors commented that magnetic microparticles coated with streptavidin is not only included in Roche assay kits but also available from Sigma-Aldrich Corporation. The streptavidin microparticle suspension (0.72 ng/mL streptavidin concentration) in Hepes (2-[4-(2-hydroxyethyl)piperazin-1-yl]ethanesulfonic acid)-bovine serum albumin buffer (pH 7.4) has a biotin-binding capacity of 1.34 μmol/L of free biotin. The authors centrifuged the streptavidin-coated microparticle suspension and removed the supernatant and then incubated serum/plasma specimen with streptavidin-coated microparticles for 1 h at room temperature with gentle shaking to deplete biotin from the specimens (5:1 ratio of streptavidin reagent to serum; N5 protocol). Following incubation the specimen was centrifuged for 10 min and then the supernatant was collected for analysis. The authors reported that their protocol effectively removed biotin from sera in all patients with multiple sclerosis taking 300 mg supplement, as the serum biotin level was found to be below the level of quantitation using liquid chromatography combined with tandem mass spectrometry. However, two healthy volunteers showed low level of biotin in specimens collected 2 h after ingestion of biotin (27.6 and 8.3 ng/mL) after neutralization procedure. Most hormone test results using Roche assays were initially abnormal but were normalized after N5 protocol. The authors concluded that most streptavidin-biotin hormone immunoassays are affected by high biotin concentrations in serum leading to a risk of misdiagnosis, but their N5 protocol is effective in suppressing biotin interference.[19]

CONCLUSIONS

Although biotin interferences in biotin-based assays are relatively new findings, the use of high-dose biotin supplement (5–10 mg per day) for hair and nail growth as well as pharmaceutical dosage of biotin (30–300 mg) may increase in the future. Therefore clinical scientists, pathologists, physicians, and healthcare professional must be aware of such effects on clinical laboratory test results using certain biotin-based immunoassays. However, it is also important to note individuals not taking any biotin supplement as well as people taking multivitamins (biotin concentration usually 30 μg, the daily recommended intake) or biotin supplements (500 μg to 2.5 mg) should have serum biotin concentrations significantly below threshold concentrations needed for significant clinical interference (more than 20% change in value from baseline). Therefore biotin interference is expected in relatively few specimens compared with the total number of specimens analyzed in any clinical laboratory. At this point, it is difficult to assess what percentage of specimens is affected because biotin interference is observed only in people taking high-dose supplement. Usually taking 5–10 mg biotin supplement daily affects tests most susceptible to biotin interferences, but all tests may be affected in people taking 100–300 mg biotin daily under medical supervision. Therefore it is very important for clinicians to ask patients about their potential use of dietary supplement including biotin.

The most commonly reported clinical problem of biotin interference is misdiagnosis of hyperthyroidism or Graves disease because of falsely decreased TSH value and falsely elevated FT_4 and FT_3 values. Although repeated testing to resolve this issue is possible by asking a patient to stop taking biotin and then retesting after 2–3 days, for acute situations such option is not available. Moreover, missed diagnosis of myocardial infarction due to falsely decreased troponin I and/or troponin T values is also a very serious problem.

Major problem of biotin interference is positive interference in competitive immunoassays but negative interferences in sandwich immunoassays. In typical competitive assays for small molecules such as FT_4, FT_3, testosterone, estradiol, and cortisol, biotin interferences falsely increase analyte values but in assays that use sandwich format (for example, TSH, Tg, FSH, LH, insulin, and autoantibodies), values are falsely lowered. This combination of two types of biotin interferences can create a significant clinical confusion, for example, high steroid hormone concentrations with suppressed LH or FSH may falsely suggest presence of a tumor. Even a slight change to results can pose serious ramifications for tests used in the diagnosis of serious infectious agents such as HIV or hepatitis C virus. The practical and immediate solution to overcome biotin interference will be to increase patient awareness of biotin interference in laboratory tests. Physicians need to counsel patients to stop taking biotin for a few days before drawing blood. With biotin interference so pervasive, it is critical that laboratorians and clinicians are aware of biotin interference so that misdiagnosis and inappropriate treatment can be prevented.[57]

REFERENCES

1. Trueb RM. Serum biotin levels in women complaining of hair loss. *Int J Trichol.* 2016;8:73–77.
2. Yates AA, Schlicker SA, Suitor CW. Dietary reference intakes: the new basis for recommendations for calcium and related nutrients. *J Am Diet Assoc.* 1998;98:699–706.
3. Donaldson M, Touger-Decker R. Vitamin and mineral supplements: friend or foe when combined with medications? *J Am Dent Assoc.* 2014;145:1153–1158.
4. Kamangar F, Emadi a. Vitamin and mineral supplements: do we really need them? *Int J Prev Med.* 2012;3:221–226.
5. Mock DM. Biotin: from nutrition to therapeutics. *J Nutr.* 2017;147:1487–1492.
6. Samarsinghe S, Meah F, Singh V, Basit A, et al. Biotin interference with routine clinical immunoassays: understanding the causes and mitigate risk. *Endocr Pract.* 2017;23:989–998.
7. Wonderling R. *A Closer Look at the Recent FDA Safety Communication about Biotin Interference.* Medical Laboratory Observer; March 2018. https://www.mlo-online.com/closer-look-recent-fda-safety-communication-biotin-interference.
8. Biotin (vitamin B7): Safety communication-May interfere with lab tests (issued by the FDA). https://www.fda.gov/safety/medwatch/safetyinformation/safetyalertsforhumanmedicalproducts/ucm586641.htm.
9. Piketty ML, Souberbielle JC. Biotin an emerging analytical interference. *Ann Biol Clin.* 2017;75:366–368.
10. Zempleni J, Wijerante SS, Hassan Y. *Biotin. Biofactors.* 2009;35:36–46.
11. Livaniou E, Evangelatos GP, Ithakissios DS, Yatzidis H, et al. Serum biotin levels in patients undergoing chronic hemodialysis. *Nephron.* 1987;46:331–332.
12. Watanabe T, Yasumura S, Shibata H, Fukui T. Biotin status and its correlation with other biochemical parameters in the elderly people of Japan. *J Am Coll Nutr.* 1998;17:48–53.
13. Clevidence BA, Marshall MW, Canary JJ. Biotin levels in plasma and urine of healthy adults consuming physiological doses of biotin. *Nutr Res.* 1988;8:1109–1118.
14. Mock DM, Mock N. Serum concentrations of bisnorbiotin and biotin sulfoxide increase during both acute and chronic biotin supplementation. *J Lab Clin Med.* 1997;129:384–388.
15. Trambas C, Lu Z, Yen T, Silkaris. Characterization of the scope and magnitude of biotin interference in susceptible Roche Elecsys competitive and sandwich immunoassay. *Ann Clin Biochem.* 2018;55:205–215.
16. Mardach R, Zempleni J, Wolf B, Cannon MJ, et al. Biotin dependency due to a defect in biotin transport. *J Clin Invest.* 2002;109:1617–1623.
17. Grimsey P, Frey N, Bending G, Zitzler J, et al. Population pharmacokinetics of exogenous biotin and the relationship between biotin serum levels and in vitro immunoassays interference. *Int J Pharmacokinetic.* 2017;2:247–256.
18. Peyro Saint Paul L, Debruyne D, Bernard D, Mock DM, Defer GL. Pharmacokinetics and pharmacodynamics of MD1003 (high-dose biotin) in the treatment of progressive multiple sclerosis. *Expert Opin Drug Metab Toxicol.* 2016;12:327–344.
19. Piketty ML, Prie D, Sedel F, Bernard D, et al. High-dose biotin therapy leading to false biochemical endocrine profiles: validation of a simple method to overcome biotin interference. *Clin Chem Lab Med.* 2017;55:817–825.
20. Sathyanarayana Rao TS, Christopher R, Andrade C. Biotin supplements and laboratory test results in neuropsychiatric practice and research. *Indian J Psychiat.* 2017;59:405–406.
21. Gounden V, Sacks DB, Zhao Z. Interference of cerebrospinal fluid total protein measurement by povidone-iodine contamination. *Clin Chim Acta.* 2015;440:3–5.
22. Zempleni J, Mock JM. Bioavailability of biotin given orally to humans in pharmacological dosage. *Am J Clin Nutr.* 1999;69:504–508.

23. Henry JG, Sobki S, Arafat N. Interference by biotin on measurement of TSH and FT4 by enzyme immunoassay on Boehringer Mannheim ES700 analyser. *Ann Clin Biochem*. 1996;33(pt. 2):162–163.

24. Kwok JS, Chan IH, Chan MH. Biotin interference on TSH and free thyroid hormone measurement. *Pathology*. 2012;44:278–280.

25. Wijerantne NG, Doery JC, Zx L. Positive and negative interference in immunoassays following biotin ingestion: a pharmacokinetic study. *Pathology*. 2012;44:674–675.

26. Al-Salameh A, Becquemont L, Brailly-Tabard S, Aubourg P, et al. A somewhat bizarre case of Graves' disease due to vitamin treatment. *J Endocr Soc*. 2017;1:431–435.

27. De Roeck Y, Philipse E, Twickler TB, Van Gaal L. Misdiagnosis of Graves' hyperthyroidism due to therapeutic biotin intervention. *Acta Clin Belg*. 2017;3:1–5.

28. Elston MS, Sehgal S, Du Toit S, Yarndley T, et al. Factitious Graves' disease due to biotin Immunoassay interference-a case and review of the literature. *J Clin Endocrinol Metab*. 2016;101:3251–3255.

29. Kummer S, Hermsen D, Distelmaier F. Biotin treatment mimicking Graves' disease. *N Engl J Med*. 2016;375:704–706.

30. Biscolla RPM, Chiamolera M, Kanashiro I, Maciel RMB. A single 10 mg dose of biotin interferes with thyroid function tests. *Thyroid*. 2017;27:1099–1100.

31. Barbesino G. Misdiagnosis of Graves' disease with apparent severe hyperthyroidism in a patient taking biotin megadoses. *Thyroid*. 2016;26:860–863.

32. Rulander NJ, Cardamome D, Senior M, Snyder PJ, et al. Interference from anti-streptavidin antibody. *Arch Pathol Lab Med*. 2013;137:1141–1146.

33. Berth M, Willaert S, De Ridder C. Anti-streptavidin IgG antibody interference in anti-cyclic citrullinated peptide (CCP) IgG antibody assays is a rare but important cause of false positive anti-CCP results. *Clin Chem Lab Med*. 2018;56:1263–1268.

34. Trambas CM, Sikaris KA, Lu ZX. More on biotin treatment mimicking Graves' disease. *N Engl J Med*. 2016;375:1698.

35. Piketty ML, Polak M, Flechtner I, Gonzales-Briceño L, et al. False biochemical diagnosis of hyperthyroidism in streptavidin-biotin-based immunoassays: the problem of biotin intake and related interferences. *Clin Chem Lab Med*. 2017;55:780–788.

36. Waghray A, Milas M, Nyalakonda K, Siperstein AE. Falsely low parathyroid hormone secondary to biotin interference: a case series. *Endocr Pract*. 2013;19:451–455.

37. Meany DL, Jan de Beur SM, Bill MJ, Sokoll LJ. A case of renal osteodystrophy with unexpected serum intact parathyroid hormone concentrations. *Clin Chem*. 2009;55:1737–1741.

38. Ranaivosoa MK, Ganel S, Agin A, Romain S, et al. Chronic kidney failure and biotin: a combination inducing unusual results in thyroid and parathyroid investigations, report of 2 cases. *Nephrol Ther*. 2017;13:553–558. [article in French].

39. Willeman T, Casez O, Faure P, Gauchez AS. Evaluation of biotin interference on immunoassays: new data for troponin I, digoxin, NT-Pro-BNP, and progesterone. *Clin Chem Lab Med*. 2017;55:e216–e219.

40. Ali M, Rajapakshe D, Cao L, Devaraj S. Discordant analytical results caused by biotin interference on diagnostics immunoassays in a pediatric hospital. *Ann Clin Lab Sci*. 2017;47:638–640.

41. Collinson PO, Gaze D, Goodacre S. The clinical and diagnostic performance characteristics of the high sensitivity Abbott cardiac troponin I assay. *Clin Biochem*. 2015;48:275–281.

42. Li D, Radulescu A, Sherstha RT, Root M, et al. Association of biotin ingestion with performance of hormone and non-hormone assays in healthy adults. *J Am Med Assoc*. 2017;318:1150–1160.

43. Rodriguez JJ, Acosta F, Bourgeois L, Dasgupta A. Biotin at high concentration interferes with the LOCI digoxin assay but the PETINIA phenytoin assay is not affected. *Ann Clin Lab Sci*. 2018;48:164–167.

44. Williams GR, Cervinski MA, Nerenz RD. Assessment of biotin interference with qualitative point of care hCG test devices. *Clin Biochem*. 2018;53:168–170.

45. Clerico A, Plebani M. Biotin interference on immunoassay methods: sporadic cases or hidden epidemic? *Clin Chem Lab Med*. 2017;55:777–779.

46. Holmes EW, Samarasinghe S, Emanuele MA, Meah F. Biotin interference in clinical immunoassays: a cause for concern. *Arch Pathol Lab Med*. 2017;141. 1459–1450.

47. Roche Diagnostics: Understanding biotin lab interference: Biotin facts. http://biotinfacts.roche.com/understand/.

48. Lipner SR, Scher RK. Biotin for the treatment of nail disease: what is the evidence? *J Dermatolog Treat*. 2017;9:1–4.

49. Paxton A. Beauty fad's ugly downside: test interference. *CAP Today*. September 20, 2016:1–9.

50. Batista MC, Ferreira CES, Faulhaber ACL, Hidal JT, et al. Biotin interference in immunoassays mimicking subclinical Graves' disease and hyperestrogenism: a case series. *Clin Chem Lab Med*. 2017;55:e99–e103.

51. Aguirre JJ, Yao J, Stier T, Theel E. Biotin interference with biotin-streptavidin based VITROS hepatitis Be antigen (HBE ag) and hepatitis Be antibody (Anti-HBe) immunoassays. *Am J Clin Pathol*. 2018;149(suppl 1):S23–S23.

52. Lim SK, Pilon A, Guechot J. Biotin interferes with free thyroid hormone and thyroglobulin, but not TSH measurements using Beckman-Access. *Ann Endocrinol*. 2017;78:186–187.

53. Selby C. Interference in immunoassay. *Ann Clin Biochem*. 1999;36:704–721.

54. Ross DS, Burch HB, Cooper DS, Greenlee MC, et al. 2016 American Thyroid Association Guidelines for diagnosis and management of hyperthyroidism and other causes of thyrotoxicosis. *Thyroid*. 2016;26:1343–1421.

55. Minkovsky A, Lee MN, Dowlatshahi M, Angell TE, et al. High dose biotin treatment for secondary progressive multiple sclerosis may interfere with thyroid assays. *AACE Clin Case Rep*. 2016;2:e370–e373.

56. Sulaiman RA. Biotin treatment causing erroneous immunoassay results: a word of caution for clinicians. *Drug Dis-
cov Ther*. 2016;10:338–339.
57. Lipner SR. Rethinking biotin therapy for hair, nail and
skin disorders. *J Am Acad Dermatol*. 2018;78:1236–1238.
58. Trambas C, Lu Z, Yen T, Sikaris K. Depletion of biotin using
streptavidin-coated microparticles: a validated solution to
the problem of biotin interference in streptavidin-biotin
immunoassays. *Ann Clin Biochem*. 2018;55:216–226.

FURTHER READING

1. Chun KY. Biotin interference in diagnostic tests. *Clin
Chem*. 2017;63:619–620.

CHAPTER 5

Issues of Interferences in Clinical Chemistry Tests Including Heterophilic Antibody Interferences

INTRODUCTION

Clinical laboratories play an important part in overall patient care. Although laboratory services may represent approximately 5% of a hospital's budget, laboratory test results account for 60%–70% of all critical decision-making by clinicians and healthcare professionals, such as admittance, discharge, and medication.[1] The First Report of the House of Commons Select Committee on Health published on May 1, 2002, states that up to 70% of all diagnosis in National Health Services (NHS, UK) patients depend on laboratory test results. Therefore NHS pathology services are critical for day-to-day evidence-based care of patients. Laboratory test results also contribute to 80% of all objective data in the clinical record and influence 60%–70% of critical decision-making.[2]

Immunoassays are widely used in clinical chemistry laboratory along with other analytic methods, such as ion selective electrodes for measuring various electrolytes in serum or plasma, atomic absorption for measuring heavy metals, enzymatic assays, and chromatographic methods. Analytic methods for analysis of common analytes in clinical chemistry laboratory (including common tests in toxicology) are listed in Table 5.1. In most hospitals, there is no separate toxicology laboratory; therefore, common tests in toxicology such as alcohol measurement and urine drug screen are conducted in the clinical chemistry laboratory. This chapter focuses on issues of interferences in immunoassays used in clinical chemistry laboratory. Although interferences commonly falsely increase analyte values (positive interference), negative interferences where analyte values are falsely lower have also been reported. In Chapter 6, issues of interferences in immunoassays used for therapeutic drug monitoring are discussed, while in Chapter 7, issues of interferences in drugs of abuse and toxicology testing are addressed.

SOURCES OF ERRORS IN CLINICAL LABORATORY TESTS

Errors in clinical laboratory test results may occur preanalytically, during analysis, or postanalytically. In one study the authors reported that among a total of 51,746 analyses, clinicians notified the authors of 393 questionable findings, 160 of which were confirmed as laboratory errors. Of the 160 confirmed errors, 61.9% were preanalytic errors, 15% were analytic, and 23.1% were postanalytic.[3] Plebani and Carraro commented that most of the laboratory mistakes (74%) did not affect patients' outcome. However, in 19% patients, laboratory mistakes were associated with further inappropriate investigations, thus resulting in an unjustifiable increase in costs. Moreover, in 6.4% patients, laboratory mistakes were associated with inappropriate care or inappropriate modification of therapy.[4] Medical errors can be traditionally clustered into four categories: errors of diagnosis, errors of treatment, errors of prevention, and an "other miscellaneous" category. Laboratory errors can affect any of these four categories of medical errors and thus are a serious hazard to patient health.[5]

Effect of Age, Gender, Diet, and Exercise on Laboratory Test Results

It is also important to remember that many factors such as age, gender, diet, exercise, and ethnicity/race impact clinical laboratory test results, which are not laboratory errors. These factors can be classified into two categories: nonmodifiable (age, gender, and ethnicity/race) and modifiable (diet, exercise). Reference intervals have been established, in some cases, to account for the effects of age and gender that are easily displayed by modern electronic laboratory information systems. On average, albumin, calcium, magnesium, hemoglobin, ferritin, and iron concentrations are lower in females.[6] Mean serum creatinine levels are usually

TABLE 5.1
Analytic Methods for the Analysis of Common Analytes in Clinical Chemistry Laboratory

Analyte	Method of Analysis
Acetaminophen[a]	Enzymatic colorimetric method or immunoassays
Alcohol[a]	Serum/plasma alcohol can be measured by enzymatic assay (based on alcohol dehydrogenase) or headspace gas chromatography
ALT	ALT is measured as activity using a specific substrate such as L-alanine
Albumin	Dye binding method using bromocresol green or bromocresol purple
ALP	ALP converts 4-nitrophenol phosphate into 4-nitrophenol, which absorbs at 405 nm
Ammonia	Enzymatic assay using glutamate dehydrogenase or ion-selective electrode
Amikacin[a]	Immunoassay
Amylase	A specific starch, for example, 2-chloro-4-nitrophenyl-α-D-maltotrioside, is used as the substrate
AST	AST is measured as activity by using a specific substrate such as L-aspartate
Bilirubin	Diazo method (colorimetric) or enzymatic methods
Brain natriuretic peptide or N-terminal prohormone brain natriuretic peptide	Specific immunoassay
Blood urea nitrogen	Chemical method (chemical reaction where urea condenses with diacetyl, forming the chromophore diazine that absorbs at 540 nm) or enzymatic method
Calcium	Spectrophotometry (arsenazo III or o-cresolphthalein complex) or ion-selective electrode
Carbamazepine[a]	Immunoassay
Carcinoembryonic antigen	Immunoassay
Chloride	Ion-selective electrode
Cholesterol	Enzymatic assay (colorimetry)
Carbon dioxide (bicarbonate)	Enzymatic method or converting plasma carbon dioxide into gaseous form prior to measurement
Cortisol	Immunoassay
CK	Enzymatic assay using phosphocreatine as substrate
CK-MB	Immunoassay
Creatinine	Creatinine is measured by Jaffe reaction (using alkaline picrate) or enzymatic assays that are less prone to assay interferences
C-reactive protein	Immunoassay
Cyclosporine	Immunoassay
Digoxin[a]	Immunoassay
Ferritin	Immunoassay
Folate	Immunoassay
FT_3	Immunoassay
FT_4	Immunoassay
Everolimus[a]	Immunoassay
Gentamycin[a]	Immunoassay
γ-GGT	GGT is measured as enzymatic activity by using a specific substrate such as γ-glutamyl-3-carboxy-4-nitroanilide

TABLE 5.1
Analytic Methods for the Analysis of Common Analytes in Clinical Chemistry Laboratory—cont'd

Analyte	Method of Analysis
Glucose	Enzymatic methods using glucose oxidase or glucose hexokinase or glucose dehydrogenase
High-density lipoprotein cholesterol	Selective protein precipitation followed by measuring cholesterol content in supernatant
Human chorionic gonado-tropin	Immunoassay
Homocysteine	Immunoassay
Lactic acid	Lactic acid is converted into pyruvic acid by LDH enzyme and NAD is converted to NADH (absorbs at 340 nm) in this reaction
LDH	Lactate is used as the substrate for LDH assay
Low-density lipoprotein cholesterol	Immunoassay
Lipase	Turbidimetric method where lipase hydrolyzes fat in solution, thus reducing turbidity
Lithium[a]	Ion-selective electrode, spectrophotometric, or enzyme assay
Magnesium	Spectrophotometric method or occasionally atomic absorption
Mycophenolic acid[a]	Immunoassay
Phenobarbital[a]	Immunoassay
Phenytoin[a]	Immunoassay
Procainamide and its active metabolite NAPA[a]	Specific immunoassay for procainamide and specific immunoassay for NAPA
Phosphorus (inorganic phosphate)	Using ammonium molybdate to form phosphomolybdate complex, which absorbs UV at 340 nm, or reducing this complex to molybdenum blue (600 nm absorption)
Potassium	Ion-selective electrode
Prostate-specific antigen	Immunoassay
Rheumatoid factor	Immunoassay
Salicylate[a]	Colorimetric (Trinder) or immunoassay
Sodium	Ion-selective electrode
Sirolimus[a]	Immunoassay
Tacrolimus[a]	Immunoassay
Theophylline[a]	Immunoassay
Tobramycin[a]	Immunoassay
Total protein	Biuret method (colorimetry) is commonly used
Transferrin	Immunoassay
Triglyceride	Enzymatic method using lipase
Troponin I	Immunoassay
Troponin T	Immunoassay
T_4	Immunoassay
T_3	Immunoassay
Thyroid-stimulating hor-mone	Immunoassay

Continued

TABLE 5.1	
Analytic Methods for the Analysis of Common Analytes in Clinical Chemistry Laboratory—cont'd	
Analyte	**Method of Analysis**
Uric acid	Phosphotungstic acid or uricase method
Urine drug screen	Specific immunoassays for individual drug screened
Vancomycin[a]	Immunoassay
Valproic acid[a]	Immunoassay
Vitamin B$_{12}$	Immunoassay

[a]In large hospital laboratories and academic medical center, these analyte tests are performed in the toxicology laboratory.
ALP, alkaline phosphatase; *ALT*, alanine aminotransferase; *AST*, aspartate aminotransferase; *CK*, creatine kinase; *FT$_3$*, free triiodothyronine; *FT$_4$*, free thyroxine; *LDH*, lactate dehydrogenase; *NAPA*, N-acetylprocainamide; *T$_3$*, triiodothyronine; *T$_4$*, thyroxine; *γ-GGT*, glutamyltransferase.

lower in females than males. Jones et al. investigated serum creatinine levels in 18,723 participants in the United States. The mean serum creatinine value was 0.96 mg/dL for women and 1.16 mg/dL for men. Interestingly, mean creatinine levels were highest in African Americans (women, 1.01 mg/dL; men, 1.25 mg/dL), lower in non-Hispanic whites (women, 0.97 mg/dL; men, 1.16 mg/dL), and lowest in Mexican-Americans (women, 0.86 mg/dL; men, 1.07 mg/dL).[7] Interestingly, cystatin C concentrations, another marker of kidney function, are also commonly lower in adolescent females than in adolescent males. The authors also observed lower serum creatinine levels in females than males in the study.[8]

Aldolase concentrations are higher in males after puberty. In one study the authors observed no difference in alkaline phosphatase (ALP) levels between boys and girls until puberty. However, higher ALP levels were noted at 10–11 years in girls and at 12–13, 14–15, and 16–17 years in boys. However, a decline in ALP concentrations begins after age 12 years for girls and 14 years for boys.[9] Menopausal women have higher ALP concentrations than males. Serum bilirubin concentrations are lower in women owing to the decreased hemoglobin concentrations. Creatine kinase (CK or total CK) concentrations are higher in males than in females. However, CK levels are increased after exercise.[10] Long-distance running can also increase total concentration of CK, CK-MB, and CK-BB.[11] Cardiac troponin I (cTnI) and cardiac troponin T (cTnT) levels are also increased after endurance exercise in athletes, but elevated N-terminal prohormone brain natriuretic peptide (NT-proBNP) concentration after exercise was not related to exercise-induced increases in cTnI or cTnT levels, but correlated with exercise time. The authors concluded that increases in NT-proBNP levels can be found in a major part of obviously healthy athletes after prolonged

TABLE 5.2	
Effect of Exercise on Common Analytes	
Analytes With Increased Value After Exercise	**Analytes With Decreased Value After Exercise**
• Alkaline phosphatase	• Activated partial
• Alanine aminotransferase	thromboplastin time
• Aspartate aminotransferase	• Fibrinogen
• Bilirubin	• Magnesium
• Brain natriuretic peptide[a]	• Platelets
• Bblood urea nitrogen	
• Creatinine	
• CK total	
• CK-MB[a]	
• D-dimer	
• Lactate dehydrogenase	
• NT-proBNP[a]	
• Myoglobin	
• Prothrombin time	
• Platelet count	
• Troponin I[a]	
• White blood cell count	

[a]Increases are significant in athletes engaged in very strenuous exercises.
CK, creatine kinase; *NT-proBNP*, N-terminal prohormone brain natriuretic peptide.

strenuous exercise. However, the release of BNP during and after exercise may not result from myocardial damage but may have cytoprotective and growth-regulating effects.[12] Effects of exercise on various commonly measured analytes are summarized in Table 5.2. Foran et al.[13] reviewed the effects of exercise on laboratory tests.

Females have higher albumin concentrations than males of the same age.[14] Lipid profiles are also affected by gender. In general, females, under the age of 20 years, have higher total cholesterol concentrations than males, but between the ages of 20 and 45 years, males commonly have higher total cholesterol concentrations than females. Males experience a peak in lipid concentrations generally between the ages of 40 and 60 years, whereas peak lipid concentrations occur between the ages of 60 and 80 years in females. Men also have higher 24-h urinary excretions of epinephrine, norepinephrine, cortisol, and creatinine excretion than women.[15] Women have higher serum γ-glutamyltransferase (GGT) and copper levels and higher reticulocyte count (due to increased erythrocyte turnover) than men.

Diet has a marked effect on laboratory test results as noted by changes in lipid profiles and glucose measurements. These changes reflect the metabolic and endocrine processes associated with food composition, amount, and intake intervals. With the exception of estimated glomerular filtration rate calculation, a few tests have widely used ethnic specific reference intervals.

Preanalytic, Analytic, and Postanalytic Errors

Most common cause of preanalytic errors include specimen mislabeling, improper specimen collection tube, specimen collection tube not filled properly, and test request errors. Moreover, improper handling of specimens, transport, and storage may also affect laboratory test results.[16] In one study the authors reported that improper request, incorrect timing of sample, wrong tube collection, and in vitro hemolysis of samples amounted to the major proportion of errors.[17] Each laboratory has specimen rejection criteria to reduce erroneous results related to preanalytic errors in laboratory tests. Moreover, laboratories also implement quality assurance programs to improve overall quality of the laboratory. Nevertheless, specimen rejection still occurs in clinical laboratories. In one study the authors reported that out of 32,548 samples received during a 7-month period in the laboratory, 177 samples (0.54%) were rejected. The most common reasons for rejection in hematologic and biochemical specimens were clotted blood specimen (51.2%), improperly labeled specimen (14.46%), and hemolyzed blood samples (11.45%). For microbiology, causes of rejection included labeling errors, collection of specimen in wrong containers, specimen collection date and time not being entered, unacceptable specimen source, and delayed transit time (18.2% each).[18] In another report based on 971,780 biological specimens submitted to the laboratory over a 1-year period, 26,070 (2.7%) specimens were rejected. The most frequent reason for rejection was clotted specimen (55.8% of total rejections), followed by inadequate volume (29.3% of total rejections). Most of the clotted specimens were received from adult hospital inpatient services (54.3%), followed by pediatric hospital inpatient services (26.8%). High rates of inadequate volume were also observed in samples originating from adult and pediatric hospital inpatient services, especially in the premature, neonatal, intensive care, and oncology units.[19]

Dikmen et al. reported that 27,067 specimens out of 453,171 samples sent to the laboratory during a 12-month period were rejected. The rejection rate was 2.5% for biochemical tests, 3.2% for complete blood count (CBC), 9.8% for blood gases, 9.2% for urine analysis, 13.3% for coagulation tests, 12.8% for therapeutic drug monitoring, 3.5% for cardiac markers, and 12% for hormone tests. The most frequent causes of rejection were fibrin clots (28%) and inadequate volume (9%) for biochemical tests, clotted samples (35%) and inadequate volume (13%) for coagulation tests, blood gas analyses, and CBC.[20] Turner et al.[21] reported that the introduction of electronic requesting in primary care can reduce the number of preanalytic errors and can improve the quality of information received with each request.

Although analytic errors are lower in magnitude than preanalytic and postanalytic errors, unlike other errors, analytic errors are often unpredictable despite considerable research in this field. A good example is biotin interferences in immunoassays that utilize biotinylated antibody and streptavidin-coated microparticles to capture biotinylated antibody. Although normal biotin level does not cause any interference, elevated biotin levels may cause significant underestimation of analyte values using sandwich immunoassay format but overestimation of analyte values using competitive assay format. This book focuses on analytic errors caused by biotin and other interfering substances.

Postanalytic errors are also a significant cause of inaccurate laboratory test results. Postanalytic errors include incomplete report, delay in reporting results, critical laboratory test result not called, incorrect reporting of result, and result reported to wrong provider. Nevertheless, postanalytic errors are less in magnitude than preanalytic errors due to widespread adoption of laboratory automation and interfaced laboratory reporting. Quality monitors for the postanalytic process emphasize critical result notification, meeting established turnaround time goals, and review of changed reports.[22] In order to reduce pre- and postanalytic errors, accreditation bodies such as the Joint Commission International and the College of American Pathologists now require clear and effective procedures for patient/sample identification and communication of critical results.[23]

INTERFERENCES DUE TO CROSS-REACTIVITY

Presence of cross-reactive substances in the specimen is a major cause of assay interference but such interferences are more common in therapeutic drug monitoring (please see Chapter 6) and drugs of abuse screen in urine using immunoassays (please see Chapter 7). Nevertheless, such interferences are also encountered in other immunoassays, for example, spironolactone interferes in the immunoassay of androstenedione.[24]

Krasowski et al. investigated issues of cross-reactivities in Roche Elecsys assays for cortisol, dehydroepiandrosterone (DHEA) sulfate, estradiol, progesterone, and testosterone and reported that the cortisol and testosterone II assays showed a wide range of cross-reactivity when compared with DHEA sulfate, estradiol II, and progesterone II assays. For the Roche cortisol assay, six compounds showed cross-reactivity of greater than 5%, while 17 additional compounds showed cross-reactivity between 0.5% and 4.9% at a test concentration of 1000 ng/mL. Therefore patients

taking prednisolone and 6-methylprednisolone may show interference with Roche cortisol immunoassay causing falsely elevated values. False elevation in cortisol levels may also occur in patients with 21-hydroxylase deficiency due to elevated 21-deoxycortisol in serum. Several anabolic steroids cross-react with Roche testosterone II assay at a concentration of 100 ng/mL. Seven compounds showed cross-reactivity of 5% or greater, while nine additional compounds showed cross-reactivity of 0.5%–4.9%. However, anabolic steroids such as oxymetholone, stanozolol, and Turinabol showed no cross-reactivity with the Roche testosterone II assay. Compounds with cross-reactivity more than 5% in Roche Elecsys assays for cortisol, progesterone, and testosterone are listed in Table 5.3. Although exemestane showed only 0.09% cross-reactivity with Roche progesterone assay, possible assay interference is expected from exemestane if progesterone is measured during peak exemestane level, which may be as high as 441 ng/mL. No compound showed more than 5% cross-reactivity in DHEA sulfate or

TABLE 5.3
Compounds With More Than 5% Cross-Reactivity With Roche Elecsys Assays for Cortisol, Progesterone, and Testosterone Assays

Analyte	Cross-Reactive Substance	Cross-Reactivity (%)	Comments
Cortisol	6-Methylpredisolone	249	Patients taking this drug may show falsely elevated cortisol levels
	Prednisolone	148	Patients taking this drug may show falsely elevated cortisol levels
	Allotetrahydrocortisol	165	Effect on cortisol assay is not characterized due to unknown serum level
	21-Deoxycortisol	45.4	Falsely elevated cortisol level only in patients with 21-hydroxylase deficiency
	Fludrocortisone	7.7	Unlikely to affect cortisol assay due to low serum level
Progesterone	5β-Dihydroprogesterone	18.2	Possible effect in patients with progesterone concentration at the lower end of reference range
Testosterone	Methyltestosterone		May falsely elevate testosterone levels
	Boldenone	7.2	Low risk of interference
	19-Norclostebol	6.7	Effect unknown
	Norethindrone	6.7	Possible interference
	11-β-Hydroxytestosterone	5.5	Effect unknown
	Methandrostenolone	5.4	Effect unknown
	Normethandrolone	5.4	Effect unknown

Source of data: Crowley RK, Broderick D, O'Shea T, Boran G, et al. Spironolactone interference in the immunoassay of androstenedione in a patient with a cortisol secreting adrenal adenoma. *Clin Endocrinol.* 2014;81:629–630.

estradiol assay.[25] Interference in urine cortisol measurement due to components present in vaginal contraceptive ring (NuvaRing) has been reported.[26] Interference from 25-hydroxyvitamin D and other circulating metabolites of vitamin D in immunoassays for vitamin D has also been reported.[27]

ISSUES OF MACROENZYMES

Macroenzymes are high-molecular-weight versions of specific serum enzymes formed either by complex formation with a high-molecular-weight serum component such as immunoglobulin or by self-polymerization. Macroenzymes frequently interfere with enzyme assays thus falsely increasing serum enzyme levels. In 1967, Berk et al. first reported the presence of macromolecular form of amylase in sera of patients with persistently elevated serum amylase levels. These patients had essentially normal renal functions.[28] Later reports of macromolecular forms of various enzymes appeared in the medical literature, including macro-lactate dehydrogenase (macro-LDH), macro-alkaline phosphatase (macro-ALP), macro-aspartate aminotransferase (macro-AST), macro-alanine aminotransferase (macro-ALT), macro-CK, macro-γ-GGT, macro-acid phosphatase, macro-lipase, and macro-leucine aminopeptidase. Macroenzymes are usually discovered in patients who show persistent elevation of a particular serum enzyme but have no symptom. Macroenzymes can be broadly classified under two categories: immunoglobulin-bound enzyme or non–immunoglobulin-bound enzyme. Immunoglobulin-bound enzymes could be considered as a specific antigen-antibody complex. Macroenzymes are mostly benign in nature because many patients with macroenzymes are healthy. However, the presence of macroenzymes in many pathologic conditions, such as certain infections and parasitic disease, neoplasm, immune disorders, and endocrine disorders, has been reported. Various methods including polyethylene glycol (PEG) precipitation, gel filtration chromatography (GFC), and serum protein electrophoresis can be used to detect the presence of macroenzyme in a patient's sera.[29,30] Commonly encountered macroenzymes are listed in Table 5.4.

Various macroenzymes are usually encountered in patients over 60 years of age. However, macro-AST is more frequent in subjects under the age of 60 years, particularly in females. Cases with this abnormality have also been reported in children, suggesting that macro-AST may start early in life but is unlikely to be congenital.[31,32]

Case Report

A 31-year-old woman was admitted to the authors' institution for elevated serum AST activity of unknown origin fluctuating between 60 and 450 IU/L (normal value, 10–42 IU/L) over the previous 4 months. The patient was admitted four times in the previous 9 years for causes unrelated to liver dysfunctions. On admission the patient did not complain of any symptoms. She denied alcohol intake or use of hepatotoxic drugs or drugs potentially affecting liver enzyme activity and also had no contact with any individual suffering from any type of hepatitis. Her physical examination was unremarkable with no jaundice, hepatosplenomegaly, or stigmata of chronic liver disease. Biochemical analysis showed a three-to fourfold elevation of AST level, with a normal level of ALT. All other liver function test results (total bilirubin, conjugated bilirubin, ALP, γ-GGT, albumin, prothrombin time) were normal, as were routine laboratory tests. In addition, results of hepatitis A, B, and C serology; Epstein-Barr virus; cytomegalovirus; and *Toxoplasma* antibody analyses were negative. Test results for metabolic (α1-antitrypsin, ceruloplasmin, serum and urine copper) diseases were within the normal range. The unremarkable findings in the history of the patient, together with a negative

TABLE 5.4
Commonly Encountered Macroenzymes

Macroenzyme	Comments
Macro-alkaline phosphatase	Complexed with IgG or lipid aggregates
Macro-alanine aminotransferase	Complexed with IgG
Macro-aspartate aminotransferase	Complexed with IgG
Macro-γ-glutamyltransferase	Complexed with IgG or lipid aggregates
Macro-lactate dehydrogenase	Complexed with IgA
Macro-amylase	Either complexed with IgA or exists as substrate-complex
Macro-lipase	Mostly complexed with IgG
Macro-creatine kinase	Mostly complexed with IgG (macro-CK-1) or macro-CK type 2 enzymes made up of mitochondria-derived self-CK polymerization
Macro-trypsin	Protease-inhibitor complex

physical examination and normality of other liver function tests, led the authors to suspect the condition of macro-AST. The subsequent studies (immunoprecipitation and immunoelectrophoresis) of AST isoenzymes revealed that AST was combined with IgG, carrying [kappa] chains to form a macromolecular antigen-antibody (AST-IgG) complex, thus confirming the diagnosis of macro-AST. The patient's family members (both the parents and two sisters) were also studied in respect of this pathologic condition, but they all had normal levels of serum transaminase. The patient was followed up for another 44 months, during which she was found to be in good health, although she continued to present with persistent AST elevation when checked by conventional serum assay, fluctuating from 60 to 383 IU/L.[32]

Macro-CK is a neglected cause of falsely elevated serum CK levels but such increases may not have any clinical significance. Macro-CKs are macroenzymes with high molecular weight and prolonged half-life. Macro-CK type 1 enzymes are complexes formed by one of the CK isoenzymes and an immunoglobulin and are found in healthy individuals as well as in a wide range of pathologic conditions. Macro-CK type 2 enzymes are made up of mitochondria-derived CK polymers and are associated with neoplasms. Both types of macro-CK enzymes have been observed but prevalence is very low. In one report, over a 10-year period the authors observed only five cases related to macro-CK. Three patients had macro-CK type 1 and two patients had macro-CK type 2, a man with a neuroendocrine carcinoma and a woman with rheumatoid arthritis.[33]

Other Macroanalytes

Macro-thyroid-stimulating hormone (TSH) interferes with TSH immunoassays. This is discussed in the Issues of Interference in TSH Immunoassays section of this chapter. Another macroanalyte, macroprolactin, interferes with prolactin immunoassay. Macroprolactin is a large protein complex of 150 kDa or more formed due to binding of monomeric prolactin (molecular weight, 23 kDa) with immunoglobulin, predominately IgG. Macroprolactin has a slower clearance rate consistent with that of IgG, leading to accumulation in the circulation and causing falsely elevated prolactin levels in immunoassays. The prevalence of macroprolactin is 1.5%–3.7%. However, in patients who are hyperprolactinemic, the reported prevalence of macroprolactin ranges from 4% to 46% depending on the assay and referral population. Macroprolactin may not have any pathologic significance. Because there is significant overlap in the clinical presentation of patients with true hyperprolactinemia and those with macroprolactin, differentiation cannot always be made on the basis of symptoms. A lack of recognition of the presence of macroprolactin can lead to unnecessary laboratory investigations, imaging, and pharmacologic or surgical treatment.[34]

Case Report

A 35-year-old female patient complaining of secondary amenorrhea, mild obesity, hirsutism, severe headache, and blurred vision was treated with dydrogesterone (10 mg daily) with good response and the patient could maintain regular menstrual cycle for about 1 year. After that, recurrence of oligomenorrhea occurred despite twofold increase in dose. The treatment was discontinued that resulted in permanent amenorrhea and the patient was referred to author's hospital. Physical examination revealed mild hirsutism. For hormonal analysis, venous blood samples were taken in the morning, after 30 min of rest (sampling referred to the second and third admissions made during the early follicular phase of the menstrual cycle). Serum prolactin, estradiol (E2), luteinizing hormone (LH), follicle-stimulating hormone (FSH), and testosterone levels were measured by the use of commercially available kits (Beckman Coulter) with analytic sensitivity. Hormonal analysis showed normal thyroid function, but serum prolactin level was highly elevated (10,610 mIU/L) with suppressed gonadotropins levels (LH, 1.1 U/L; FSH, 1.2 U/L; E2, 235 pmol/L). PEG precipitation test revealed that hyperprolactinemia was almost exclusive owing to the presence of macroprolactin (macroprolactin level after PEG precipitation was 4.7% of the original value). An invasive pituitary macroadenoma was visualized on magnetic resonance imaging (MRI), and cabergoline therapy was initiated. Disappearance of clinical signs and symptoms, normalization of gonadotropin levels, and restoration of regular ovulatory menstrual cycles after 1 year of treatment were observed. The significant decrease in macroprolactin levels and tumor volume in response to dopamine agonist therapy suggests the tumoral origin of this isoform.[35]

HETEROPHILIC ANTIBODY INTERFERENCE

Heterophilic antibodies are endogenous proteins that bind animal antigens. The heterophilic antibodies are polyclonal and heterogeneous in nature, consisting of the following types:

- Heterophilic antibodies that interact poorly and nonspecifically with the assay antibodies.
- Anti-animal antibodies that interact strongly and specifically with the assay antibody, a common example is human antimouse antibody (HAMA).

- Endogenous human autoantibodies interfering with an assay.
- Therapeutic antibodies, where antibodies given therapeutically interfere with an assay.
- Rheumatoid factor can also be broadly classified as an heterophilic antibody.

Heterophilic antibodies may arise in a patient in response to exposure to certain animals or animal products, due to infection by bacterial or viral agents, or nonspecifically. Although many of the immunoglobulin clones in normal human serum may display anti-animal antibody properties, only those antibodies with sufficient titer and affinity toward the reagent antibody used in the assay may cause clinically significant interference. Prevalence of heterophilic antibody varied widely in various published reports. In one study the prevalence of heterophilic antibody was 0.2%–3.7%.[36] Heterophilic antibodies are found more in sick and hospitalized patients, with reported prevalence of 0.2%–15%, but some reports claimed prevalence of up to 40% in the general population. An individual can form heterophilic antibody at any time when exposed to a foreign antigen, for example, during vaccination, antibody-targeted therapies (cancer, autoimmune disorder, etc.), and antibody-targeted imaging reagents. Exposure to animals may also lead to formation of heterophilic antibody. Moreover, blood transfusion, autoimmune disease, etc. may also result in the formation of heterophilic antibodies. Although up to 40% of samples may contain heterophilic antibodies, analytically important interferences due to the presence of heterophilic antibody in serum most likely occur in 0.5%–3% specimens.[37]

Heterophilic antibodies usually interfere with sandwich immunoassays but rarely with competitive immunoassays. In sandwich assay format, assay interference could happen when heterophilic antibodies bridge the capture and detection antibody in the assay design, resulting in elevated analyte concentration. However, rarely heterophilic antibodies may cause negative interference (falsely low analyte value) by interfering with analyte binding to capture the antibody. Grasko et al. described the case of a 60-year-old female retired nurse who underwent extensive unnecessary invasive investigation for analytic interference in adrenocorticotropic hormone (ACTH) measurement using Immulite analyzer (Siemens) most likely due to the presence of heterophilic antibodies. When ACTH was analyzed by a different immunoassay (Roche Elecsys 170 analyzer), significant discordance was observed. For example, in the specimen, ACTH concentration was 5.4 pmol/L using Immulite analyzer but <0.22 pmol/L using the Roche analyzer. Interestingly, after treating the specimen with heterophilic antibody–blocking tube (HBT) containing a specific binder called heterophilic antibody–blocking agent (Scantibodies Laboratory, Santee, CA), ACTH values were increased, rather than decreased, when measured using the Immulite assay. A possible explanation for this observation is that treatment with heterophilic antibody–blocking agent removed weakly reacting heterophilic antibodies (IgG class) leaving highly reactive specific IgM class antibodies to bind more avidly with analyte antibody, resulting in a higher assay interference.[38]

In addition to interfering with ACTH assay, heterophilic antibodies may affect a wide range of laboratory tests including falsely elevated results of endocrine tests, tumor markers, cardiac markers, and very rarely therapeutic drug monitoring (digoxin, one case report, as well as cyclosporine and tacrolimus using antibody-conjugated magnetic immunoassay). However, heterophilic antibody may also cause falsely lower cortisol levels causing incorrect diagnosis of hypothalamic-pituitary-adrenal axis insufficiency. False-low levels of thyroglobulin have also been reported. However, interferences of heterophilic antibody in immunoassay of various tumor markers are particularly important because they may lead to false diagnosis of malignancy.[39] Bolland et al.[40] described a case where heterophilic antibodies cause false-low cortisol values in a 42-year-old woman. Weakly false-positive rapid human immunodeficiency virus antibody testing owing to the presence of heterophilic antibody in a 24-year-old woman has been described.[41] Bolstad et al.[42] reported that 21 assays covering 19 different analytes were susceptible to interferences due to heterophilic antibodies. Immunoassays of various analytes affected by heterophilic antibodies are listed in Table 5.5.

Rheumatoid Factor Interferences

Rheumatoid factors are IgM-type antibodies that interact with assay antibodies at the Fc area and are present in the serum of greater than 70% of patients with rheumatoid arthritis. Rheumatoid factors are also found in patients with other autoimmune diseases and their titer may also increase during infection or inflammation. Interferences of rheumatoid factors are more commonly observed in sandwich immunoassays but may also occur in competitive immunoassay format. The mechanisms are similar to interferences caused by heterophilic antibodies. In competitive type immunoassays, rheumatoid factors bind to the assay antibody, preventing its reaction to the label reagent through steric hindrance, thus generating false-positive results. If

TABLE 5.5
Interference of Heterophilic Antibodies in Various Immunoassays

Test Category	Example of Individual Tests	Comments
Cardiac markers	Troponin I, CK-MB, BNP	Usually falsely elevated levels.
Thyroid function tests	TSH, FT$_3$, FT$_4$, thyroglobulin	In general, values are falsely elevated but both falsely elevated and false-low values have been reported with thyroglobulin testing.
Hormones	ACTH, cortisol, calcitonin, FSH, LH, growth hormone, prolactin, estradiol, progesterone, testosterone, PTH	Usually falsely elevated values but negative interference in cortisol testing has been reported.
Therapeutic drugs	Digoxin, cyclosporine, tacrolimus	Falsely elevated values for all three drugs but only ACMIA cyclosporine and tacrolimus assays are affected by heterophilic antibody interference.
Tumor markers	PSA, CEA, CA 19-9, CA 125, AFP, β-hCG, calcitonin	Usually falsely elevated values, which may cause wrong diagnosis of malignancy, a very serious issue of interference.
Infectious disease	HIV testing	Weakly false-positive test result using point of care testing.
Other analytes	C-reactive protein, D-dimer, insulin, SHBG	Falsely elevated values

ACMIA, antibody-conjugated magnetic immunoassay; *ACTH*, adrenocorticotropic hormone; *AFP*, α-fetoprotein; *BNP*, B-type natriuretic peptide or brain natriuretic peptide; *CA 19-9*, cancer antigen 19-9; *CA 125*, cancer antigen 125; *CEA*, carcinoembryonic antigen; *CK-MB*, creatine kinase MB isoenzyme; *FSH*, follicle-stimulating hormone; *FT$_3$*, free triiodothyronine; *FT$_4$*, free thyroxine; *LH*, luteinizing hormone; *PSA*, prostate-specific antigen; *PTH*, parathyroid hormone; *SHBG*, sex hormone–binding globulin; *TSH*, thyroid-stimulating hormone; *β-hCG*, β-human chorionic gonadotropin.

rheumatoid factors are suspected to cause interference, the patient's history needs to be examined. It is also advisable to measure the rheumatoid factor levels in serum by using commercially available immunoassays.

ISSUES OF INTERFERENCES IN THYROID FUNCTION ASSAYS

Thyroid function tests are among the most commonly requested laboratory investigations in both primary and secondary care. Although interpretations of thyroid function tests, such as TSH, free thyroxine (FT$_4$), total thyroxine (T$_4$), free triiodothyronine (FT$_3$), and total triiodothyronine (T$_3$), in most patients are straightforward to confirm the clinical impression of euthyroidism, hypothyroidism, or hyperthyroidism, in some patients, discordance between test results and clinical picture may be observed for a variety of reasons. In such cases, it is important first to revisit the clinical context and to consider potential confounding factors, such as alterations in normal physiology (e.g., pregnancy), intercurrent (nonthyroidal) illness, and medication usage (e.g., T$_4$, amiodarone, and heparin). Once these have been excluded, laboratory artefacts in commonly used TSH or other thyroid function tests using

immunoassays should be considered as the source of discordance. Ruling our laboratory errors before treating the patient is essential to avoid unnecessary further investigation and/or treatment.[43]

Issues of Interference in Thyroid-Stimulating Hormone Immunoassays

Most commercially available TSH assays are based on sandwich format where two different antibodies, each recognizing a part of the TSH molecule, are used. Biotin interferences in TSH immunoassays that utilize biotinylated antibodies have been discussed in detail in Chapter 4. In this chapter, other sources of interferences in TSH immunoassays are discussed. In general, macro-TSH, heterophilic antibodies, and rheumatoid factors are known to interfere with TSH immunoassays.

Macro-TSH is a large molecule that is a complex of TSH with immunoglobulins, most commonly IgG. Macro-TSH interferes with all TSH immunoassays causing elevated serum TSH levels but has no effect on T$_4$ levels. As a result, patients with macro-TSH have elevated serum TSH and normal FT$_4$ levels, thus mimicking subclinical hypothyroidism. In one study the authors screened for macro-TSH in specimens collected from 1901 patients with subclinical hypothyroidism using the PEG method and confirmation with GFC. The authors reported that the prevalence of macro-TSH was only 0.79% (15/1901). Commercial immunoassay systems variably recognized macro-TSH but the Architect TSH immunoassay (Abbott Laboratories) showed lower interference than the Elecsys (Roche) and ADVIA Centaur (Siemens) TSH assays, but still recognized 60% of specimens containing macro-TSH.[44] In another study the authors reported that 11 out of 681 patients were diagnosed with subclinical hypothyroidism (1.62%) because of the interference of macro-TSH.[45]

Case Report

A 60-year-old man was admitted to the hospital for a left intertrochanteric fracture after an episode of syncope. His medical history included hypertension, ischemic heart disease, and dyslipidemia. His oral medications were aspirin, atenolol, enalapril, lovastatin, and sublingual glyceryl trinitrate. The thyroid function tests showed a markedly elevated TSH level (232 mIU/L; reference interval, 0.45–5.0 mIU/L) with a low-normal FT_4 (10 pmol/L; reference interval, 10.0–23.0 pmol/L) using Vitros 5600 analyzer and immunoassays (Ortho Clinical Diagnostics). However, the patient showed no symptom of hypothyroidism. The antithyroid peroxidase antibody level was elevated at 496 IU/mL (reference interval, 0–50 IU/mL) but no elevation was observed in the levels of antithyroglobulin and anti-TSH receptor antibodies. Falsely elevated TSH level was suspected and reanalysis of the specimen using ADVIA Centaur TSH assay (Siemens) showed a discordant value of 122 mIU/L. However, FT_4 value was comparable. Dilution of specimen showed nonlinearity, indicating assay interference. The serum of the patient was then subjected to PEG precipitation to remove high-molecular-weight proteins, including interfering antibodies and macro-TSH. The post-PEG TSH recovery for this patient was 3.2% (pre-PEG, 122 mIU/L; post-PEG, 3.9 mIU/L using the ADVIA Centaur assay), indicating the presence of a high-molecular-weight interfering substance (macro-TSH or heterophilic antibody or rheumatoid factor). However, HBT studies and rheumatoid factors were negative indicating that neither heterophilic antibody nor rheumatoid factor was responsible for assay interference but mixing the patient's sample with a hypothyroid patient sample showed reduced TSH recovery. GFC demonstrated a TSH peak fraction that approximated the molecular size of IgG; together with the excess TSH-binding capacity. The authors concluded that the interference was due the presence of TSH-IgG macro-TSH. A review of 12 macro-TSH case reports showed

that samples with macro-TSH produce overrecovery with dilution, return negative results on anti-animal and anti-heterophile blocking studies, and commonly have recovery of less than 20% when subjected to PEG precipitation.[46] Interferences of heterophilic antibody and rheumatoid factors in TSH immunoassays have also been reported.[47]

Case Report

A patient with moderate suspicion of thyroid disease showed an elevated TSH value of 119 mIU/L (reference range, 1.5–3.5 mIU/L) but normal FT_4 level (17.1 pmol/L) using Roche modular analyzer. When the measurement was repeated in another laboratory using Abbott Architect analyzer, the TSH value was significantly lower at 25.2 mIU/L. When the specimen was distributed to other laboratories, TSH value was 8.8 mIU/L using the ADVIA Centaur assay, 40 mIU/L using the Beckman Coulter assay, 62 mIU/L using the Immulite assay (Siemens), and 95 IU/L using the Vitros analyzer (Ortho Clinical Diagnosis). The TSH value was reduced to the normal value of 2.7 mIU/L after removal of IgG, by protein G affinity chromatography, indicating that TSH value was falsely elevated due to interference of heterophilic antibody.[48]

Issues of Interference in Other Thyroid Function Tests

Heterophilic antibodies may falsely increase serum FT_4 levels. In one study the authors reported two patients who presented with nonspecific symptoms of tiredness that prompted their clinician to order a thyroid function test. In both cases, FT_4 level was reported to be high, with a normal TSH value. Repeated blood tests from the same laboratory using the same assay method yielded similar results but when tests were repeated in a different laboratory, both TSH and FT_4 levels were normal. The authors suggested that heterophilic antibodies were interfering with the FT_4 assay, leading to clinical confusion in both patients.[49]

Falsely elevated TSH values, along with other thyroid functions test results, in a patient due to the interference of heterophilic antibody have also been reported. A 57-year-old man with a history of liver fibrosis, portal hypertension, and hypersplenism had thyroid function assessed because of weight loss. His FT_4, total T_4, total T_3, and TSH assayed by enhanced chemiluminescent immunoassay showed elevated values but the patient was euthyroid. Suspecting interference, patient' serum was reanalyzed following immunoglobulin precipitation with 50% PEG. The levels of T_4 and T_3 were then found to be within the reference ranges, suggesting that the serum contained interfering immunoglobulins. Anti-immunoglobulin antibodies also neutralized the interfering substances present in the patient's serum. These observations indicate that nonspecific immunoglobulins present in the patient's serum cross-reacted with the antibodies employed in the assays used causing falsely elevated results. The authors speculated that heterophilic antibodies present in patient's serum may cause this interference.[50]

False-positive thyroglobulin values due to the presence of rheumatoid factors has been reported.[51] However, both false low and falsely elevated thyroglobulin levels due to interference of heterophilic antibodies have also been reported. Giovanella and Ghelfo[52] reported that heterophilic antibodies may cause negative interference in thyroglobulin measurement. In contrast, Ding et al.[53] reported positive interference of heterophilic antibodies in thyroglobulin measurement using Immulite 2000 analyzer (Siemens).

Thyroid hormone autoantibodies (THAAs) are endogenously produced antibodies developed against diiodothyronine, T_3, or T_4. These autoantibodies may interfere with the measurement of thyroid hormones but the magnitude of interference may differ depending on the specificity of assay antibodies. THAAs are present most likely in less than 2% of the general population and in approximately 10% of patients with thyroid as well as nonthyroid autoimmune diseases. However, much higher percentage (approximately 40%) of patients with autoimmune thyroid diseases may have THAAs. Despite the high prevalence of THAAs in some patient population, interference of these antibodies in routine measurement of thyroid hormones is relatively low, most likely less than 2%.[47] The pathophysiology of THAAs is poorly understood because these autoantibodies are not only associated with patients with autoimmune thyroid disease such as Hashimoto thyroiditis and Graves disease but also associated with interferon therapy, Epstein-Barr virus infection, Sjögren syndrome, rheumatoid arthritis, lupus, and others. Sakata et al.[54] reported that the prevalence of anti-T_3 or anti-T_4 antibodies among healthy population could be as high as 1.8%.

Interference of THAAs, for example, anti-T_4 antibody, in one-step immunoassay for FT_4 is more significant than in two-step immunoassay. In one study, FT_4 measured by one-step chemiluminescent assay (ADVIA Centaur XP system, Siemens) in a 40-year-old female with Hashimoto disease was 54.70 pmol/L (reference range, 11.46–23.17 pmol/L) but when measured by a two-step immunoassay (Architect i2000 analyzer, Abbott Laboratories), the FT_4 value was 13.00 pmol/L, a value within the reference range. However, total T_3

(1.50 nmol/L by Siemens and 1.55 nmol/L by Abbott assay) and TSH (6.23 mIU/L by Siemens and 5.97 mIU/L by Abbott assay) values did not differ when analyzed by different methods. The authors commented that while one-step immunoassay is vulnerable to THAAs, which directly compete with endogenous FT_4, a two-step assay utilizing an intermediate washing step induces a noncompetitive reaction that removes the unbound FT_4 and interfering factor. Therefore no interference was observed using the Abbott two-step immunoassays.[55]

Case Report
A full-term female infant was born to a 34-year-old mother with no medical history. The initial cord blood was submitted for routine neonatal congenital hypothyroidism screening that showed TSH level of 20.2 mIU/L and FT_4 of 31.1 pmol/L. Suspecting interference, a fresh venous specimen was analyzed 5 days later, which showed normal TSH level of 2.52 mIU/L but elevated FT_4 level of 46.7 pmol/L using Immulite 2000 analyzer (Siemens). When aliquots of cord blood and venous blood were subjected to PEG precipitation and reanalyzed, FT_4 value in the cord blood was reduced to 15.3 pmol/L and FT_4 value in the venous blood was also reduced to 27.7 pmol/L, confirming assay interference. In addition, when a second venous specimen collected 11 days after the original cord blood sample was analyzed using cobas analyzer (Roche Diagnostics), the TSH level was 2.05 mIU/L and FT_4 was 21.4 pmol/L. The authors speculated that analysis of FT_4 using a one-step immunoassay such as Immulite is subjected to more interference from anti-thyroid hormone antibodies than a two-step immunoassay such as Roche assay. The authors also suggested that gradual decline in FT_4 values indicated that the presence of anti-thyroid antibodies in the neonate was due to the transplacental transfer from mother to fetus because newborns, due to their immature immune system, are unlikely to produce anti-thyroid hormone antibodies.[56]

Serum thyroglobulin measurement is routinely conducted as a postoperative tumor marker test for patients with differentiated thyroid cancer. However, thyroglobulin immunometric assays are prone to interference. Heterophilic antibodies including HAMAs are known to falsely increase serum thyroglobulin levels. In contrast, thyroglobulin antibodies cause a false low or undetectable serum thyroglobulin level that can have more serious consequences because it can mask the presence of disease.[57] It has been estimated that endogenous thyroglobulin antibodies are present in up to 25% of patients with differentiated thyroid cancer.[58]

Calcitonin, a hormone produced by C-cells in the thyroid gland, counteracts the action of the parathyroid hormone and plays an important role in regulating calcium and phosphate levels in blood. In one study the authors investigated heterophilic antibody interference in serum calcitonin assay by studying 378 patients with thyroid nodules shown not to have medullary thyroid carcinoma after extensive diagnostic workup. Serum calcitonin measurement was performed using Immulite 2000 analyzer (Siemens) before and after incubating each serum sample in HBTs and the differences were calculated. The authors identified 5 out of 378 patients studied (1.3%) who had heterophilic antibody interference in calcitonin assay. However, only four out of five patients demonstrated clinically significant false-positive calcitonin results. The authors recommended that serum pretreatment in HBTs should be considered when increased serum calcitonin levels are found in a patient with a thyroid nodule to prevent unwarranted investigations or therapies.[59]

ISSUES OF INTERFERENCE IN HORMONE ANALYSIS

As mentioned earlier, the presence of heterophilic antibodies may falsely increase ACTH and calcitonin values but may falsely decrease cortisol values. In addition, heterophilic antibodies may also affect thyroid hormone analysis. However, heterophilic antibodies may interfere with other hormone analysis. Cheng et al. reported the case of a previously healthy postmenopausal woman who presented with a falsely elevated total testosterone level due to interference consistent with heterophilic antibodies. Her testosterone level was normal when analyzed by liquid chromatography-mass spectrometry.[60] Webster et al. described two cases of immunoglobulin interference in serum FSH assays using Immulite 2000 analyzer (Siemens) illustrating two common mechanisms for false-positive interference in two-site (sandwich) immunoassays. The first case describes a circulating autoimmune FSH immunoglobulin complex (macro-FSH), and the second case, a cross-linking antibody directed against the assay reagents. Immunoglobulin interference was detected and characterized using a combination of method comparison, immunosubtraction, and size exclusion chromatography.[61] A low serum level of insulinlike growth factor 1 (IGF-1) is considered as a diagnostic indicator of growth hormone deficiency. Negative interference of heterophilic antibody in IGF-1 measurement using immunoassay has been reported.

Case Report

A 56-year-old male patient was referred to the outpatient clinic with a medical history of decline in physical performance, forgetfulness, unexplained osteopenia, and weight loss (about 3 kg over the past 8 months). The patient had no history of exposure to animal proteins, but in his childhood, he had lived on a farm and was exposed to diverse animal species. Although clinical symptoms were unremarkable, laboratory test results indicated possibility of secondary hypothyroidism (FT_4, 3.9 pmol/L; measured by Immulite 2500 assay, Siemens) and secondary hypogonadism (testosterone, <0.70 nmol/L). Moreover, serum level of IGF-1 (total IGF-1; also measured by Immulite, 9.5 nmol/L) was at the lower end of reference range, but IGF-1 bioactivity measured by the kinase receptor activation assay was normal. Two growth hormone stimulation tests were performed, but their results did not support growth hormone deficiency. MRI of the brain demonstrated a normal pituitary gland with no abnormalities. Interference of heterophilic antibody was considered, and after specimen pretreatment with HBTs (Scantibodies Laboratory Inc), FT_4 (20.0 pmol/L), testosterone (22.0 nmol/L), and IGF-1 (16.1 nmol/L) values returned to normal levels, thus confirming negative interference of heterophilic antibodies in FT_4, testosterone, and IGF-1 measurements using immunoassays. In contrast the authors observed positive interference of heterophilic antibodies in FSH assay (FSH, 3.0 U/L before treatment but 1.9 U/L after treatment with heterophilic antibody–blocking agent).[62]

ISSUES OF INTERFERENCES IN TUMOR MARKER ASSAYS

Interferences from heterophilic antibodies are a serious problem in immunoassays of various tumor markers. Heterophilic antibodies can be of IgG, IgA, IgM, or rarely IgE class. Sandwich immunoassays are particularly prone to interference to IgG class heterophilic antibodies. In addition, rheumatoid factors can also interfere with tumor marker assays. Moreover, other factors can cause significant interferences with measurement of specific tumor markers. Commonly and less commonly measured tumor markers in clinical laboratories are summarized in Table 5.6.

The prostate-specific enzyme is a marker for prostate cancer, and the major cause of false elevation of prostate-specific antigen (PSA) value is the presence of heterophilic antibody in the serum.[63] Kamiyam et al. described a case of a 63-year-old man who was diagnosed with prostate cancer with a PSA level of 5.27 ng/mL.

TABLE 5.6
Commonly and Less Commonly Measured Tumor Markers

Commonly Measured Tumor Markers	Less Commonly Measured Tumor Markers
• Prostate-specific antigen • Human chorionic gonadotropin (also measured for detecting pregnancy) • α-Fetoprotein • β2-Microglobulin • CA 125 • CA 19-9 • CA 15-3 • CA 27.29 • Carcinoembryonic antigen • Estrogen receptor/progesterone receptor • Thyroglobulin	• CA 72-4 • CYFRA 21-1 (cytokeratin fragment) • Human epididymis protein (HE4) • Human epidermal growth factor receptor 2 (HER2/neu) • Squamous cell carcinoma antigen • Chromogranin A • *BRCA1* and *BRCA2* gene mutation

CA, cancer antigen or carbohydrate antigen.

Radical prostatectomy was performed but his PSA level remained elevated at 3.32, 4.78, and 5.93 ng/mL at 1, 2, and 3 months, respectively, after surgery. However, radiologic investigation revealed no metastasis. The authors concluded that PSA values in this patient were falsely elevated due to interference from heterophilic antibodies.[64] Park et al. reported a case of HAMA interference with a commonly used PSA assay, leading to falsely elevated levels (2.17–2.46 ng/mL) after radical prostatectomy using ADVIA Centaur assay (Bayer Diagnostics, now Siemens). Preoperative PSA level was 4.4 ng/mL. An alternate PSA assay using goat detection antibody (Hybritech PSA assay, Beckman Coulter) eliminated interference, with all values around 0.05 ng/mL. The authors concluded that when a patient's PSA level is inconsistent with the clinical scenario, interference of HAMA in serum PSA assay must be considered.[65]

Cancer antigen 125 or carbohydrate antigen 125 (CA 125) is a marker for ovarian cancer. Meigs syndrome is the association of ovarian fibroma, pleural effusion, and ascites, which may also cause marked elevation of CA 125. Measurable CA 125 concentrations can also be observed in patients without any cancer. CA 125 concentrations are known to rise in patients with severe congestive heart failure, and such elevations correlate with the severity of disease and increased concentration of brain natriuretic peptide (BNP), a marker for heart failure.[66] Sometimes F(ab')2 fragments of the murine monoclonal antibody (OC 125) are administered to patients with ovarian cancer because OC 125 is directed against the CA 125 antigen present on the surface of human ovarian cancer cells. Exposure to such antibody may lead to development of an immune response causing the presence of HAMA, which may interfere in an unpredictable manner with the determination of CA 125 using serum specimens in such patients.[67] Berthlof et al.[68] also reported falsely elevated CA 125 levels using immunoassays due to interference of heterophilic antibodies.

CA 19-9, also called sialylated Lewis (a) antigen, is a tumor marker used primarily in the management of pancreatic cancer. CA 19-9 can be elevated in many types of gastrointestinal cancer, such as colorectal cancer, esophageal cancer, and hepatocellular carcinoma (HCC). Apart from cancer, elevated levels may also occur in pancreatitis, cirrhosis, and diseases of the bile ducts. It can be elevated in people with obstruction of the bile duct.[69] Although interferences in CA 19-9 assays are less frequently reported, in one report a 61-year-old Caucasian male suffering from fatigue and weight loss showed an elevated CA 19-9 level of 80 kU/L using ADVIA Centaur analyzer (Siemens). Determination of CA 19-9 on Vidas, AxSYM, and Architect i2000 systems gave normal results. His rheumatoid factor concentration was very high (900 kIU/L). The authors commented that although interferences in the CA 19-9 assay are not frequent, the ADVIA Centaur system appears to be more sensitive to rheumatoid factor interference than other immunoassays.[70]

Case Report

A 45-year-old male presented to his primary care physician in early 2007 with constant tiredness; diarrhea, two to three times per day; and dyspepsia. His medical history included Perthes disease in childhood (right hip), which recurred 6–7 years ago, and pleurisy. He was on atorvastatin for hypercholesterolemia and sertraline hydrochloride for anxiety. Incidental testing gave a CA 19-9 result of 659 kU/L and a repeat about a month later was >700 kU/L (reference range, <41 kU/L using ADVIA Centaur, Siemens). He was referred to a gastroenterologist for further investigation including pancreatic cancer but other cancer markers were normal except for CA 19-9, which was highly elevated at 868 kU/L. A repeat of the CA 19-9 test using Roche Modular E170 analyzer showed a normal CA 19-9 result of 16 kU/L. A sample from this patient was finally sent to a third hospital that used a different assay for CA 19-9 (Brahms KRYPTOR) and the result was found to be 8 kU/L (reference range, <33 kU/L). The authors suspected assay interference by heterophilic antibodies but treating specimens with HBTs did not eliminate such interferences. Based on further studies, the authors identified a novel interference, a non-CA 19-9, low-molecular-weight entity, causing falsely elevated CA 19-9 result using ADVIA Centaur analyzer.[71]

Serum α-fetoprotein (AFP) is the most commonly used marker for diagnosis of HCC because its level is often elevated in patients with HCC. Serum levels of AFP do not correlate well with other clinical features such as size, stage, or prognosis. False-positive elevations of serum AFP concentrations can occur from tumors of the gastrointestinal tract, or from liver damage (e.g., cirrhosis, hepatitis, or drug or alcohol abuse).[72] As expected, heterophilic antibodies, if present in the specimen, can also cause falsely elevate AFP concentration. AFP level may also be falsely elevated due to the presence of rheumatoid factor in the specimen.[73] Wang et al.[74] reported negative interference of rheumatoid factor in the chemiluminescent microparticle immunoassay of AFP using Architect analyzer (Abbott Laboratories, Abbott Park, IL).

For the diagnosis of multiple myeloma, the serum β2-microglobulin level is one of the prognostic factors incorporated into the International Staging System. However, β2-microglobulin levels also rise with worsening renal dysfunction. Ibsen et al. reported elevated β2-microglobulin concentration in serum of children suffering from infectious mononucleosis. The diagnosis was established by demonstration of Epstein-Barr virus–specific IgM antibodies.[75]

Human chorionic gonadotropin (hCG) is a hormone composed of α- and β-subunits, and the β-subunit (β-hCG) provides functional specificity to this hormone. hCG is synthesized in large amounts by placental trophoblastic tissue and in much smaller amounts by the hypophysis and other organs such as testicles, liver, and colon. Therefore elevated levels of hCG are observed during pregnancy, produced by the developing placenta after conception, and later by the placental component syncytiotrophoblast.[76] Although major application of hCG is in the initial diagnosis of pregnancy, it is also used as a cancer marker for testicular and ovarian cancers.

A positive test for hCG before menopause is an indication of pregnancy. However, positive hCG test in a postmenopausal woman is a diagnostic challenge. Erroneous assumption of cancer due to positive hCG may lead to costly diagnostic tests and even unnecessary cancer chemotherapy in the absence of cancer. Pituitary is capable of producing hCG. Low levels of hCG accompany midcycle preovulatory surge of LH and production of hCG by pituitary also continues after menopause. Therefore low level of hCG in serum is normal in postmenopausal women (level up to 32 mIU/mL reported).[77] In addition, low levels of hCG test result due to pituitary hCG is also found in women who have a history of gestational trophoblastic disease.

For a positive hCG serum level, it is important to determine if the hCG represents an actual early pregnancy (intrauterine or ectopic), active gestational trophoblastic disease (complete or partial mole, invasive mole, choriocarcinoma, placental site trophoblastic tumor, etc.), quiescent gestational trophoblastic disease, a laboratory false-positive result (also called phantom hCG), or a physiologic artifact (pituitary hCG). De Backer et al. reported persistent low hCG levels in two nonpregnant women. In the first case, hCG levels raised during several years following a spontaneous abortion. The likelihood of false-positive results due to interference of heterophilic antibodies was ruled out and extensive clinical investigation excluded the presence of a tumor. The diagnosis was quiescent gestational trophoblastic disease. In the second case, elevated hCG level was an incidental finding where the clinician questioned the laboratory regarding validity of the test. The cause was probably an increase in pituitary hCG production and chronic renal failure in the patient.[78]

Phantom hCG and phantom choriocarcinoma syndrome (pseudohypergonadotropinemia) refer to persistent mildly elevated hCG levels, which may mislead physicians to treat patients with cytotoxic chemotherapy for choriocarcinoma when in reality no true hCG level or trophoblastic disease is present. In one patient with low levels of serum hCG (49–89 mIU/mL) 11 months after the miscarriage, diagnosis of choriocarcinoma was made. Despite chemotherapy and a hysterectomy, low levels of hCG were still detected in serum using three different assays, but no hCG was detected in urine. When serum was diluted, levels did not decrease parallel to dilution, indicating that the molecule measured in her serum was a pseudogonadotropin or phantom hCG, an interfering substance in the test. In the second case, positive hCG was observed 7 years after normal pregnancy in a woman. A pelvic ultrasound and a laparoscopy revealed no pregnancy but hCG concentration stayed between 48 and 74 mIU/mL over a 3-month period. Although low levels of hCG, β-hCG, and β-core fragment were detected in serum, no activity was observed in urine. Therefore hCG in serum was due to phantom hCG. The authors also described a third case where false-positive hCG level in serum (51–135 mIU/mL) was due to phantom hCG.[79]

Persistent low levels of phantom hCG may also be due to heterophilic mouse antibody interference, which is a serious problem in analysis of hCG in serum. However, heterophilic antibodies are absent in urine due to their high molecular weight. Therefore urinary β-hCG test is not affected by heterophilic antibodies. Positive serum hCG in the absence of urinary β-hCG is a strong indicator of interferences in serum hCG measurements due to the presence of heterophilic antibodies. In one report the authors investigated 12 women who were diagnosed with postgestational choriocarcinoma on the basis of persistently positive hCG test results in the absence of pregnancy. Specimens collected from these women were tested for hCG, hCG-free β-subunit, and hCG β-core fragment. False-positive hCG concentrations were identified by two criteria: (1) detection of hCG in serum and lack of detection of hCG and its degradation products in urine and (2) wide variations in results for different serum hCG assays. The authors further corroborated false-positive hCG values by the lack of parallel changes in hCG results when serum was diluted, by false detection of other antigens, and by failure to detect hCG with in-house assays. All 12 women met both criteria for false-positive hCG, and all had corroborating findings. In all 12 cases, a false diagnosis had been made, and most of the women had been subjected to needless surgery or chemotherapy. Assay kinetics indicated that heterophilic antibodies were responsible for the false-positive results. As a result, further therapy was stopped in these women. The authors concluded that current protocols for the diagnosis and treatment of choriocarcinoma should be modified to include a compulsory test for hCG in urine.[80]

ISSUES OF INTERFERENCES IN CARDIAC MARKER ASSAYS

cTnI, cTnT, and CK-MB are frequently measured in clinical laboratories as a marker for cardiac damage such as myocardial infarction. BNP or NT-proBNP are also measured as markers of heart failure. Interference of macro-CK in measurement of total CK has been discussed earlier in this chapter. However, macro-CK can also interfere with CK-MB immunoassays. Axinte et al. commented that when assessing an acute coronary syndrome by means of high serum levels of CK-MB, one must keep in mind that there are several other causes for an increase in the level of these markers, such as myocarditis, pericarditis, heart failure, severe aortic stenosis, stroke, renal failure, malignant hyperthermia, Reye syndrome, polymyositis, and borreliosis. In addition, macro-CK may also falsely elevate serum CK-MB levels. The authors reported two clinical cases where macro-CK was the cause of apparent increase in serum

CK and CK-MB: in a 79-year-old male with a history of coronary disease and an 82-year-old female with permanent atrial fibrillation.[81] Sztefko et al. reported case of an 82-year-old female patient who was admitted to the emergency department after a fainting episode, low blood pressure, and arrhythmia. The patient reported a watery diarrhea for 3 weeks. She showed extremely high total CK (CK, 478 U/L first measurement) and CK-MB (948 U/L in first measurement) activity but normal troponin level (<0.01 ng/mL). Laboratory investigation showed the presence of macro-CK. Abnormal CK and CK-MB activity was still measured 8 months after the initial episode. Due to unexplained watery diarrhea during 3 weeks before admission, several tests for the presence of autoantibodies were performed. The only positive result has been noted for anti-*Saccharomyces cerevisiae* antibodies, IgG isotype.[82] Macro-troponin I as well as macro-troponin T can falsely increase troponin I or troponin T values.

Case Report

A 35-year-old female presented with complaints of tiredness, shortness of breath, and recent onset of chest pain. The cTnI concentration was elevated at 6.4 ng/mL (Abbott Architect reference value, <0.03 ng/mL) with a normal electrocardiogram (ECG). Physical examination, radiologic imaging, and routine laboratory investigations did not provide an explanation for the elevated cTnI concentration. Suspecting assay interference, the authors reanalyzed specimens with different cTnI assays (ADVIA Centaur, Siemens and Biomerieux Vidas cTnI, and Beckman Coulter Access AccuTnI) as well as measured troponin T using Elecsys 2010 (Roche) but all methods gave results below the detection limit of respective assays. Serial dilutions of sample and addition of mouse serum did not alter the results. However, PEG precipitation and GFC showed the presence of a high-molecular-weight immunoreactive protein. Using GFC and protein-A IgG precipitation, the interference could be identified as a macrocomplex containing IgG and (fragments of) cTnI. Interestingly, during a 2-year follow-up, cTnI value slowly decreased to 3.3 ng/mL.[83]

Circulating macro-troponin I may also cause elevated high-sensitive troponin I values with the Architect High Sensitive Troponin-I assay (Abbott Laboratories), with the potential to be clinically misleading.[84] Circulating macro-troponin may also affect cTnT analysis. Elevated ALP can interfere with contemporary, alkaline phosphate-dependent immunoassays (which use ALP for signal amplification) for measuring troponin I, including DxI-cTnI and DxI-hCG (Beckman Coulter). The validation of such methods should include evaluations for endogenous ALP interference, especially when ALP levels exceed 1000 U/L.[85] In addition to autoantibodies, rheumatoid factor, heterophilic antibodies, excess fibrin, and hemolysis can also falsely increase troponin levels.[86] False-positive cTnT levels (E170 analyzer, Roche Diagnostics) due to interference of heterophilic antibodies have also been reported.[87]

Case Report

A 57-year-old male with history of heavy alcohol use, tobacco use, hypertension, hyperlipidemia, and squamous cell malignancy of the tongue status post chemotherapy and radiation presented to the emergency department with complaint of bilateral lower extremity and scrotal edema with increased abdominal girth for 1 month. The patient denied any fevers, chills, nausea, vomiting, abdominal pain or discomfort, altered sensorium, diarrhea, or hematemesis. His physical examination was significant for cachexia, no scleral icterus, nonjaundiced, positive caput medusae, tense nontender abdominal ascites with shifting dullness, hepatomegaly, and resting tremor of hands without asterixis, atrophic arm musculature, lack of hair on distal lower extremities, 3+tense lower extremity edema to just below the patellae with 2+pedal pulses, and 2+scrotal edema. The patient denied any exposure to mice, rodents, or other animals. The ECG revealed sinus tachycardia at 101 beats per minute, slightly prolonged QTc 0.448, low-voltage QRS without right ventricular strain pattern, and nonspecific reduced T-wave amplitude in the anterior leads compared with 1 month prior. The patient also showed an elevated cTnI value of 41.0 ng/mL (Beckman Coulter analyzer), which remained elevated throughout his hospital course but patient had no clinical symptom indicative of acute myocardial injury. Dilution of patient's serum showed nonlinearity indicative of assay interference. The patient's sample was sent to the reference Beckman Coulter Laboratory in Chaska, MN, where treatment of the specimen with heterophilic antibody resolved the interference (cTnI value after treatment, 1.04 ng/mL), indicating that the false-positive cTnI result was due to heterophilic antibody incidence.[86]

Macro-BNP may also falsely elevate serum BNP levels. A 61-year-old female presented with nontypical chest pain showed a high level of BNP (3188 ng/L; reference range <100 ng/L, method Abbott Architect) but without any symptom of heart failure. Echocardiography did not provide an explanation for the elevated BNP concentrations. In follow-up, the chest pain complaints disappeared but BNP remained elevated at the same levels. Samples were also sent to other laboratories for measurement of BNP with the Siemens ADVIA

Centaur and Beckman Coulter Unicel DxI and for the measurement of NT-proBNP with the Siemens Immulite 2000 and Siemens Dimension Vista. All these assay methods gave results that were within the reference values of the appropriate assay. Interestingly, PEG precipitation showed the presence of a high-molecular-weight immunoreactive protein. The authors concluded that false-positive BNP result was possibly caused by a macro-BNP, which was only immunoreactive in the Abbott Architect BNP immunoassay.[88]

PARAPROTEIN INTERFERENCES IN IMMUNOASSAYS

Paraproteins are monoclonal immunoglobulins or immunoglobulin fractions that are produced by a clonal population of B-cell lineage cells. The presence of paraproteins in blood or urine may be related to a benign process known as monoclonal gammopathy of unknown significance or multiple myeloma and Waldenstrom macroglobulinemia. A wide variety of laboratory tests may be affected by paraproteins, including blood counts and serum sodium, calcium, phosphorus, and high-density lipoprotein cholesterol (HDL-C) tests. Interferences could be positive or negative. Roy reported the case a 65-year-old previously healthy man with IgG kappa (3.5 g/dL) multiple myeloma with no detectable HDL-C. However, old records from his primary physician's office showed normal HDL-C level 5 and 2 years ago (46 and 42 mg/dL). The authors concluded that negative interference of paraprotein was the cause of undetectable HDL-C level in the patient.[89]

In another study the authors encountered an artifactually increased total bilirubin concentration (31.8 mg/dL) and undetectable HDL-C in a patient with a monoclonal IgM paraprotein (6.6 g/dL) using Roche total bilirubin assay and Roche HDL-C Plus assay. Patient's direct bilirubin level was 2 mg/L; total protein, 12.7 g/dL; and HDL-C, undetectable. Similar results were seen on subsequent samples (both in plasma and serum), with a high total bilirubin and undetectable HDL-C. On a previous admission 2 years before, the patient's total bilirubin concentration had been within the reference interval at 4 mg/L, and HDL-C was measured as 39 mg/dL by the same assay systems. The authors observed similar interference in other patients with paraproteins present in sera and concluded that paraprotein was responsible for elevated bilirubin but undetectable HDL-C concentration.[90] Yang et al. investigated 88 serum specimens with monoclonal gammopathies and observed that paraprotein interferences caused spuriously high levels of total bilirubin in four sera measured by the Roche Modular analyzer. In contrast, paraprotein interference on direct bilirubin was almost exclusively observed with the AU2700 analyzer (Beckman). In addition, paraprotein interferences with HDL-C results were present in 35% of specimens assayed with the Roche Modular and 16% of specimens assayed with the AU2700 analyzer.[91]

Paraprotein interferes with test results via different mechanisms. Paraprotein may interact with assay reagent, for example, slow color development was noted in the bromocresol green reaction for measuring serum albumin in a sample collected from a patient with an IgM-kappa type of paraprotein. This caused artificially low albumin value. Paraproteins, especially IgG myeloma, may also falsely lower creatinine values measured by the Jaffe reaction. A more widespread cause of paraprotein interference is due to binding of paraprotein to assay antibody, for example, false low TSH value using AxSYM analyzer (Abbott Laboratories), because paraproteins prevent binding of TSH with assay antibody. Pseudohyponatremia is a well-recognized artifact of hyperproteinemia and hyperlipidemia. Moreover, narrow anion gap is often seen in patients with paraproteinemia. Hyperphosphatasemia secondary to binding of phosphate ions with immunoglobulin, especially paraproteins, has also been reported. Macro-lipase and macro-LDH have been observed in patients with paraproteins. Paraproteins may also bind T_3 and T_4 thus falsely elevating their concentration.[92] Paraproteins may interfere with measurement of certain antibiotics.

PROZONE (HOOK) EFFECT

Falsely lower analyte concentration measured by an immunoassay in the presence of very high analyte concentration (antigen excess) due to an artifact is known as "prozone effect" of "hook effect. Characteristically, serum with high concentrations of a certain analyte can give a false negative/low result when tested using undiluted serum but shows nonlinearity during dilution.[93] The prozone effect occurs mostly (but not exclusively) in assays where all three components (antigen, antibody, and marker) are incubated simultaneously (single step assay).[94] Prozone effects are more common with assay where analyte concentrations could be very high, like C-reactive protein (100-fold increase), antistaphylolysin antibodies (10-fold increase), hormones (such as hCG), IgE (>1000-fold), ferritin (100-fold increase), and various tumor markers (especially CA 19-9, PSA) where concentrations may increase over 10,000 fold. The possibility of the hook effect is discovered by serial dilutions (1:10, 1:100, 1:1000 or even higher) where analyte values are higher in diluted specimen than the original undiluted specimen.[95]

Prozone or hook effect is commonly observed with hCG measurement in serum, In one study, the authors investigated potential hook effect in four serum hCG immunoassays using specimen from a male patient with extragonadal germ cell tumor with hCG concentration of 3,6000,000 IU/L (3.6 million IU/L). The authors observed falsely low results with ADVIA Centaur (Siemens), Immulite 2000 (Siemens) and DxI 800 (Beckman) as well as likely false negative with E170 analyzer (Roche). All these immunoassays are single-step sandwich immunoassays.[96] Hook effect may also cause false negative urine pregnancy test using point of care device.[97]

Case Report

A 27-year-old Latin-American female presented to the emergency department complaining of abdominal pain, persistent nausea, and vomiting for 2 weeks, as well as a twenty-pound weight loss. The patient reported a positive home pregnancy test 16 weeks ago but did not seek any medical advice. She thought she had miscarried 4 weeks prior, with passage of a profuse amount of "jellylike" substance, which had persisted. She had a vaginal delivery after 40 weeks of gestation 3 years ago. Ultrasound examination in the patient showed an enlarged uterus filled with echogenic material. Her serum β-hCG level was 900 IU/L. The patient was transferred to the university hospital with findings concerning for a complete molar pregnancy. Upon transfer, ultrasound findings were confirmed and serum β-hCG test was repeated, showing a value of 882 mIU/mL. The patient was found to be thrombocytopenic, anemic with a hemoglobin value of 7.5, and coagulopathic with an international normalized ratio of 2.2. After transfusion of packed red blood cells and fresh frozen plasma, the patient was taken to the operating room for suction evacuation of uterine contents. Pathology was consistent with a complete hydatidiform mole. The patient developed preeclampsia with blood pressures as high as 162/90, requiring magnesium prophylaxis. The hospital laboratory was contacted and asked to perform serial dilutions of the original β-hCG sample, reported as 882 mIU/mL. Dilutions were performed to 1:1600 and the β-hCG result was recalculated to be over 1,300,000 mIU/mL. On postoperative day 1, the β-hCG level was reduced to 326,175 mIU/mL and was further reduced to 13,035 mIU/mL on postoperative day 8. The patient recovered well and her β-hCG concentrations continued to fall, but subsequently plateaued between 456 and 489 mIU/mL after approximately 10 weeks, requiring 10 cycles of weekly methotrexate therapy after which her β-hCG value was below the detection limit of the assay.[98]

Other assays are also susceptible to prozone or hook effect. Ranjitkar et al.[99] described potential hook effect in three commercial assays for ferritin. Hook effect is also observed with PSA immunoassays. A 67-year-old man presented with a lower abdominal mass. Radiographic examination revealed a huge mass filling the entire pelvis. Although PSA level was 1.4 ng/mL, percutaneous needle biopsy revealed adenocarcinoma compatible with prostate cancer, which stained positive for PSA. Hormone therapy was initiated, and 1 month later, his PSA level was as high as 2713 ng/mL, although the mass had decreased in size. High-dose hook effect was suspected and hormone therapy was continued. Finally the patient responded to therapy and his PSA level was reduced below 0.1 ng/mL.[100]

ELIMINATING ASSAY INTERFERENCE

If a test result does not correlate with the clinical picture, false-positive or false-negative test results should be suspected. The first approach to eliminate assay interference is to retest the specimen using an immunoassay manufactured by another diagnostic company because it is extremely unlikely that two different assays for the same analyte will have similar interference from a cross-reactive substance. Another approach to eliminate assay interference is to use a different analytic method such as liquid chromatography combined with mass spectrometry or tandem mass spectrometry. However, such sophisticated analytic technique may not be available in most clinical laboratories.

Serial dilution is another approach to identify assay interference. Nonlinearity upon dilution is a strong indication of potential assay interference. Moreover, prozone or hook effect can also be identified by serial dilution and may be eliminated by using high dilution depending on the magnitude of antigen excess. Serial dilution is also useful in identifying heterophilic antibody interference in immunoassays where diluted specimen most commonly show lower result when accounted for dilution than the value observed with undiluted specimens because heterophilic antibody interference usually results in falsely elevated analyte concentration. However, heterophilic antibody may also cause negative interference, but again nonlinearity during dilution is an indication of assay interference. Various approaches to identify assay interferences are listed in Table 5.7.

TABLE 5.7
General Approaches to Identify and Overcome Assay Interference

Approach to Eliminate Assay Interference	Comments
Reanalysis of specimen using a different method	The most straightforward approach to establish assay interference is reanalysis of a specimen by a different assay platform or a more specific analytic method. This may resolve either the interference or a discordant result between two assays (>20% difference indicates assay interference).
Serial dilution	Assay nonlinearity during serial dilution is another simple way to establish assay interference. Moreover, high serial dilution (1:10 or 1:20) is effective in resolving prozone (hook) effect.
Pretreatment of specimen with heterophilic antibody–blocking agent or heterophilic antibody–blocking tubes	This procedure is effective in reducing or eliminating interferences from heterophilic antibodies but this step may not remove all interferences.
Monitoring analyte in an alternative specimen	Analyzing β-hCG in urine (heterophilic antibodies are absent in urine due to their high molecular weight) is an effective approach to remove interferences of heterophilic antibodies in serum hCG measurement.
Monitoring free analyte in the protein free ultrafiltrate	Heterophilic antibodies are absent in the protein-free ultrafiltrate due to their high molecular weight. Monitoring free digoxin eliminates interference of heterophilic antibodies in serum digoxin measurement.
Polyethylene glycol precipitation	Precipitation of interfering protein may eliminate assay interference.
Size exclusion gel chromatography	Useful in removing IgG antibody and may resolve interferences due to IgG-type heterophilic antibodies or autoantibodies.
Measuring rheumatoid factor in serum	If rheumatoid factor is suspected of causing interference, it should be measured in serum using any commercially available assay.

As described already, a patient history of any exposure to animal antibodies, illness, or exposure to animals should also alert possibility of heterophilic antibody interference. Some assay reagents incorporate "heterophilic antibody–blocking reagent." Therefore package insert of the assay should be examined carefully to see if heterophilic antibody blockers used in the assay. Nevertheless, heterophilic blocker may not be able to eliminate all interferences related to heterophilic antibodies. However, several blocking agents are commercially available: Immunoglobulin Inhibiting Reagent (IIR; Bioreclamation, NY), Heterophilic Blocking Reagent (HBR; Scantibodies, La Jolla, CA), HeteroBlock (Omega Biologicals, Bozeman, MT), and MAB 33 (monoclonal mouse IgG1) and Poly MAB 33 (polymeric monoclonal IgG1/Fab; Boehringer Mannheim). IIR is a proprietary formulation of high-affinity anti-animal antibody, and HBR is monoclonal mouse antihuman IgM. A suspected discordant sample, e.g., a sample giving false-positive hCG results, may be separately incubated with the blocker and then reassayed.[101] Reinsberg[102] studied the efficacy of various blocking reagents in eliminating HAMA interference. In another example, a clinically discordant false-positive serum myoglobin result (where another cardiac marker concentration, such as troponin I, was negative) was attributed to HAMA interference. The interference could be removed by the use of HBR.[103] It is important to note that no blocker reagent can guarantee success in all samples.

If a heterophilic antibody–blocking agent fails to eliminate such interference, another approach is to selectively remove such interfering antibodies. This can be achieved by selective adsorption of human IgG by a solid phase containing protein A or protein G. However, this step does not eliminate IgM-type interference. Alternatively, the antibody fraction in the sample may be precipitated out with a PEG reagent (preferably PEG 6000). Gessl et al. described the case of a 68-year-old woman being treated with methotrexate for rheumatoid arthritis, who showed highly elevated TSH but normal T_3 and T_4 levels using immunoassays marketed by Roche Diagnostics. Suspecting assay interference, the original specimen was diluted 1:2, 1:10, and 1:128 with Universal Diluent from Roche Diagnostics but rather acceptable linearity was observed. Treatment of specimen using heterophilic antibody–blocking agent provided by Roche did not eliminate observed interference, thus ruling out heterophilic antibody interference. Therefore an autoantibody interacting with TSH (macro-TSH) was considered as the cause of interference. The authors considered pretreatment with PEG (PEG 6000) because such pretreatment is known to remove interference of macroanalyte interference in prolactin, FSH, T_4, T3, testosterone, sex hormone–binding globulin, cortisol, and growth hormone assays. The authors observed a strikingly reduced TSH recovery (TSH reduced from 205 to 3.88 mIU/L), indicating interference in TSH assay was due to macro-TSH. The specimen was sent to Roche Research and Development laboratory in Germany and further analysis using size exclusion gel chromatography showed the presence of TSH-IgG complex. However, further testing indicated that IgG antibody interfered with ruthenium label in the TSH Roche assay. The authors concluded that anti-ruthenium antibodies mimicked macro-TSH in the electrochemiluminescent immunoassay manufactured by Roche Diagnostics. Finally, authors tested the patient's specimen using a ruthenium-independent platform (ADVIA Centaur, Siemens) and observed a normal TSH value of 1.68 mIU/L.[104]

CONCLUSIONS

When a laboratory test result does not correlate with the clinical picture of a patient, assay interference should be suspected. Assay interference may occur from both endogenous and exogenous factors. Endogenous factors such as high bilirubin, hemolysis, and lipemia may interfere with various assays. Endogenous antibodies such as autoantibodies, rheumatoid factors, and heterophilic antibodies are known to interfere with immunoassays. Similarly, exogenous factors such as a drug or drug metabolite can also cross-react with assay antibody causing erroneous results. When assay interference is suspected, reanalysis of specimen using a different assay platform may eliminate such interference or discordance between two results may indicate assay interference. Moreover, nonlinearity during dilution is also an indication of assay interference. Pretreatment of specimens with heterophilic antibody–blocking reagent and performing PEG precipitation and size exclusion gel chromatography may be useful in overcoming interferences caused by heterophilic antibodies.

REFERENCES

1. Forsman RW. Why is the laboratory an afterthought for managed care organizations? *Clin Chem*. 1996;42:813–816.
2. Hallworth MJ. The 70% claim: what is the evidence base? *Ann Clin Biochem*. 2011;48:487–488.
3. Carraro P, Plebani M. Errors in a stat laboratory: types and frequencies 10 years later. *Clin Chem*. 2007;53:1338–1342.

4. Plebani M, Carraro P. Mistakes in a stat laboratory: types and frequency. *Clin Chem*. 1997;43:1348–1351.

5. Lippi G. Governance of preanalytical variability: travelling the right path to the bright side of the moon? *Clin Chim Acta*. 2009;404:32–36.

6. Young M, Bermes EW. Specimen collection and other preanalytical variables. In: Burtis CA, ed. *Tietz Fundamentals of Clinical Chemistry*. 5th ed. Philadelphia: WB Saunders; 2001:30–54.

7. Jones CA, McQuillan GM, Kusek JW, Eberhardt MS, et al. Serum creatinine levels in the US population: third national health and nutrition examination survey. *Am J Kidney Dis*. 1998;32:992–999.

8. Groesbeck D, Kottgen A, Parekh R, Selvin E, et al. Age, gender, and race effects on cystatin C levels in US adolescents. *Clin J Am Soc Nephrol*. 2008;3:1777–1785.

9. Turan S, Topcu B, Gokce I, Guran T, et al. Serum alkaline phosphatase levels in healthy children and evaluation of alkaline phosphatase z-scores in different types of rickets. *J Clin Res Pediatr Endocrinol*. 2011;3:7–11.

10. Brancaccio P, Maffulli N, Limongelli FM. Creatine kinase monitoring in sport medicine. *Br Med Bull*. 2007;81–82:209–230.

11. Noakes TD, Kotzenberg G, McArthur PS, Dykman J. Elevated serum creatine kinase MB and creatine kinase BB-isoenzyme fractions after ultra-marathon running. *Eur J Appl Physiol Occup Physiol*. 1983;52:75–79.

12. Scharhag J, Herrmann M, Urhausen A, Haschke M, et al. Independent elevations of N-terminal pro-brain natriuretic peptide and cardiac troponins in endurance athletes after prolonged strenuous exercise. *Am Heart J*. 2005;150:1128–1134.

13. Foran SE, Lewandrowski KB, Kratz A. Effects of exercise on laboratory test results. *Lab Med*. 2003;34:736–742.

14. Thalacker-Mercer AE, Johnson CA, Yarasheski KE, Carnell NS, et al. Nutrient ingestion, protein intake, and sex, but not age, affects the albumin synthesis rate in humans. *J Nutr*. 2007;137. 1734–1340.

15. Masi CM, Rickett EM, Hawkley LC, Cacioppo JT. Gender and ethnic differences in urinary stress hormones: the population-based Chicago health, aging, and social relations study. *J Appl Physiol*. 2004;97:941–947.

16. Alsina MJ, Alvarez V, Barba N, Bullich S, et al. Pre-analytical quality control program-an overview of results (2001–2005 summary). *Clin Chem Lab Med*. 2008;46:849–854.

17. Ashakiran S, Sumati ME, Murthy NK. A study of pre-analytical variables in clinical biochemistry laboratory. *Clin Biochem*. 2011;44:944–945.

18. Bhat V, Tiwari M, Chavan P, Kelkar R. Analysis of laboratory sample rejections in the pre-analytical stage at an oncology center. *Clin Chim Acta*. 2012;413:1203–1206.

19. Sinici Lay I, Pınar A, Akbıyık F. Classification of reasons for rejection of biological specimens based on pre-preanalytical processes to identify quality indicators at a university hospital clinical laboratory in Turkey. *Clin Biochem*. 2014;47:1002–1005.

20. Dikmen ZG, Pinar A, Akbiyik F. Specimen rejection in laboratory medicine: necessary for patient safety? *Biochem Med*. 2015;25:377–385.

21. Turner HE, Deans KA, Kite A, Croal BL. The effect of electronic ordering on pre-analytical errors in primary care. *Ann Clin Biochem*. 2013;50:485–488.

22. Walz SE, Darcy TP. Patient safety & post-analytical error. *Clin Lab Med*. 2013;33:183–194.

23. Hawkins R. Managing the pre- and post-analytical phases of the total testing process. *Ann Lab Med*. 2012;32:5–16.

24. Crowley RK, Broderick D, O'Shea T, Boran G, et al. Spironolactone interference in the immunoassay of androstenedione in a patient with a cortisol secreting adrenal adenoma. *Clin Endocrinol*. 2014;81:629–630.

25. Krasowski MD, Drees D, Morris CS, Maakestad J, et al. Cross-reactivity of steroid hormone immunoassays: clinical significance and two-dimensional molecular similarity prediction. *BMC Clin Pathol*. 2014;14:33.

26. Escudero Fernández JM, Rabinovich IH, de Osaba MJM. Interference in urinary free cortisol determination by components of the NuvaRing contraceptive device. *Clin Chem Lab Med*. 2008;46:419–420.

27. Carter GD, Jones JC, Shannon J, Williams EL, et al. 25-Hydroxyvitamin D assays: potential interference from other circulating vitamin D metabolites. *J Steroid Biochem Mol Biol*. 2016;164:134–138.

28. Berk JE, Kizu H, Wilding P. Macroamylasemia: a newly recognized cause for elevated serum amylase activity. *N Engl J Med*. 1967;277:941–946.

29. Remaley AT, Wilding P. Macroenzymes: biochemical characterization, clinical significance and laboratory detection. *Clin Chem*. 1989;35:2261–2270.

30. Moriyama T, Tamura S, Nakano K, Otsuka K, et al. Laboratory and clinical features of abnormal macroenzymes found in human sera. *Biochim Biophys Acta*. 2015;1854:658–667.

31. Galasso PJ, Litin SC, O'Brien JF. The macroenzymes: a clinical review. *Mayo Clin Proc*. 1993;68:349–354.

32. Orlando R, Carbone A, Lirussi F. Macro-aspartate aminotransferase (macro-AST). A 12-year follow-up study in a young female. *Eur J Gastroenterol Hepatol*. 2003;15:1371–1373.

33. Aljuani F, Tournadre A, Cecchetti S, Soubrier M, et al. Macro-creatine kinase: a neglected cause of elevated creatine kinase. *Intern Med J*. 2015;45:457–459.

34. Samson SL, Hamrahian AH, Ezzat S. AACE Neuroendocrine and Pituitary Scientific Committee; American College of Endocrinology (ACE). American association of clinical endocrinologist, American college of endocrinology disease state clinical review: clinical relevance of macroprolactin in the absence or presence of true hyperprolactinemia. *Endocr Pract*. 2015;21:1427–1435.

35. Elenkova A, Abadzhieva Z, Genov N, Vasilev V. Macroprolactinemia in a patient with invasive macroprolactinoma: a case report and minireview. *Case Rep Endocrinol*. 2013;2013:634349.

36. Preissner CM, Dodge LA, O'Kane DJ, Singh RJ, et al. Prevalence of heterophilic antibody interference in eight automated tumor marker immunoassays. *Clin Chem.* 2005;51:208–210.

37. Preissner CM, O'Kane DJ, Singh RJ, Morris JC, et al. Phantoms in the assay tube: heterophile antibody interferences in serum thyroglobulin assays. *J Clin Endocrinol Metab.* 2003;88:3069–3074.

38. Grasko J, Willliams R, Beilin J, Glendenning P, et al. A diagnostic conundrum: heterophilic antibody interference in an adrenocorticotropic hormone immunoassay not detectable using a proprietary heterophile blocking reagent. *Ann Clin Biochem.* 2013;50:433–437.

39. Morton A. When lab tests lie: heterophilic antibody. *Aust Fam Physician.* 2014;43:391–393.

40. Bolland MJ, Chiu WW, Davidson JS, Croxson MS. Heterophile antibodies may cause falsely lowered serum cortisol values. *J Endocrinol Invest.* 2005;28:643–655.

41. Spencer DV, Nolte FS, Zhu Y. Heterophilic antibody interference causing false-positive rapid human immunodeficiency virus antibody testing. *Clin Chim Acta.* 2009;399:121–122.

42. Bolstad N, Warren DJ, Bjerner J, Kravdal G, et al. Heterophilic antibody interference in commercial immunoassays; a screening study using paired native and pre-blocked sera. *Clin Chem Lab Med.* 2011;49(12):2001–2006.

43. Koulouri O, Moran C, Halsall D, Chatterjee K, et al. Pitfalls in the measurement and interpretation of thyroid function tests. *Best Pract Res Clin Endocrinol Metab.* 2013;27:745–762.

44. Hattori N, Ishihara T, Shimatsu A. Variability in the detection of macro TSH in different immunoassay systems. *Eur J Endocrinol.* 2016;174:9–15.

45. Hattori N, Ishihara T, Yamagami K, Shimatsu A. Macro TSH in patients with subclinical hypothyroidism. *Clin Endocrinol.* 2015;83:923–930.

46. Loh TP, Kao SL, Halsall DJ, Toh SA, et al. Macrothyrotropin: a case report and review of literature. *J Clin Endocrinol Metab.* 2012;97:1823–1828.

47. Després N, Grant AM. Antibody interference in thyroid assays: a potential for clinical misinformation. *Clin Chem.* 1998;44:440–454.

48. Ross HA, Menheere PP. Endocrinology Section of SKML (Dutch Foundation for Quality Assessment in Clinical Laboratories), Thomas CM et al. Interference from heterophilic antibodies in seven current TSH assays. *Ann Clin Biochem.* 2008;45:616.

49. Ghosh S, Howlett M, Boag D, Malik I, et al. Interference in free thyroxine immunoassay. *Eur J Intern Med.* 2008;19(3):221–222.

50. Fiad TM, Duffy J, McKenna TJ. Multiple spuriously abnormal thyroid function indices due to heterophilic antibodies. *Clin Endocrinol.* 1994;41:391–395.

51. Astarita G, Gutiérrez S, Kogovsek N, Mormandi E, et al. False positive in the measurement of thyroglobulin induced by rheumatoid factor. *Clin Chim Acta.* 2015;447:43–46.

52. Giovanella L, Ghelfo A. Undetectable serum thyroglobulin due to negative interference of heterophile antibodies in relapsing thyroid carcinoma. *Clin Chem.* 2007;53:1871–1872.

53. Ding L, Shankara-Narayana N, Wood C, Ward P, et al. Markedly elevated serum thyroglobulin associated with heterophile antibodies: a cautionary tale. *Thyroid.* 2013;23:771–772.

54. Sakata S, Matsuda M, Ogawa T, Takuno H, et al. Prevalence of thyroid hormone autoantibodies in healthy subjects. *Clin Endocrinol.* 1994;41:365–370.

55. Lee MN, Lee SY, Hur KY, Park HD. Thyroxine (74) autoantibody interference of free T4 concentration measurement in a patient with Hashimoto's disease. *Ann Lab Med.* 2017;37:169–171.

56. Brincat I, Buhagiar G. The risk from anti-thyroid hormone antibody interference in neonatal congenital hypothyroidism screening. *Int J Neonatal Screen.* 2017;3:4.

57. Spencer C, Petrovic I, Fatemi S. Current thyroglobulin autoantibody (TgAb) assays often fail to detect interfering TgAb that can result in the reporting of falsely low/undetectable serum Tg IMA values for patients with differentiated thyroid cancer. *J Clin Endocrinol Metab.* 2011;96:128312–128391.

58. Spencer CA, Takeuchi M, Kazarosyan M, Wang CC, et al. Serum thyroglobulin autoantibodies: prevalence, influence on serum thyroglobulin measurement, and prognostic significance in patients with differentiated thyroid carcinoma. *J Clin Endocrinol Metab.* 1998;83:1121–1127.

59. Giovanella L, Suriano S. Spurious hypercalcitoninemia and heterophilic antibodies in patients with thyroid nodules. *Head Neck.* 2011;33:95–97.

60. Cheng I, Norian JM, Jacobson JD. Falsely elevated testosterone due to heterophile antibodies. *Obstet Gynecol.* 2012;120:455–458.

61. Webster R, Fahie-Wilson M, Barker P, Chatterjee VK, et al. Immunoglobulin interference in serum follicle-stimulating hormone assays: autoimmune and heterophilic antibody interference. *Ann Clin Biochem.* 2010;47:386–389.

62. Brugts MP, Luermans JG, Lentjes EG, van Trooyen-van Vrouwerff NJ, et al. Heterophilic antibodies may be a cause of falsely low total IGF1 levels. *Eur J Endocrinol.* 2009;161:561–565.

63. Morgan BR, Tarter TH. Serum heterophilic antibodies interfere with prostate specific antigen test and results in over treatment in a patient with prostate cancer. *J Urol.* 2001;166:2311–2312.

64. Kamiyama Y, Somiya S, Fujikawa S, Yamada Y, et al. False elevation of prostate specific antigen caused by heterophilic antibody interference after radical prostatectomy: a case report. *Hinyokika Kiyo.* 2017;63:435–437. [article in Japanese].

65. Park S, Wians Jr FH, Cadeddu JA. Spurious prostate-specific antigen (PSA) recurrence after radical prostatectomy: interference by human antimouse heterophile antibodies. *Int J Urol.* 2007;14(3):251–253.

66. Nunez J, Nunez E, Consuegra L, Sanchis J, et al. Carbohydrate enzyme 125: an emerging prognostic risk factor in acute heart failure? *Heart*. 2007;93:716–721.

67. Maher VE, Drukman SJ, Kinders RJ, Hunter RE, et al. Human antibody response to the intravenous and intraperitoneal administration of F(ab') fragment of OC 125 murine monoclonal antibody. *J Immunother*. 1991;11:56–66.

68. Bertholf RL, Johannsen L, Guy B. False elevation of serum CA-125 level caused by human anti-mouse antibodies. *Ann Clin Lab Sci*. 2002;32:414–418.

69. Goonetilleke KS, Siriwardena AK. Systematic review of carbohydrate antigen (CA 19-9) as a biochemical marker in the diagnosis of pancreatic cancer. *Eur J Surg Oncol*. April 2007;33:266–270.

70. Berth M, Bosmans E, Everaert J, Dierick J, et al. Rheumatoid factor interference in the determination of carbohydrate antigen 19-9 (CA 19-9). *Clin Chem Lab Med*. 2006;44:1137–1139.

71. Monaghan PJ, Leonard MB, Neithercut WD, Raraty MG. False positive carbohydrate antigen 19-9 (CA19-9) results due to a low-molecular weight interference in an apparently healthy male. *Clin Chim Acta*. 2009;406:41–44.

72. Javadpour N. Significance of elevated serum alpha-fetoprotein (AFP) in seminoma. *Cancer*. 1980;45:2166–2168.

73. Zhu X, Yuyingguo Huang Z, Yang S. Rheumatoid factor interference if a-fetoprotein evaluations in human serum by ELISA. *Clin Lab*. 2014;60:1795–1800.

74. Wang H, Bi X, Xu L, Li Y. Negative interference by rheumatoid factor in alpha-fetoprotein chemiluminescent microparticle immunoassay. *Ann Clin Biochem*. 2017;54:55–59.

75. Ibsen KK, Krabbe S, Hess J. Beta 2 microglobulin serum levels in infectious mononucleosis in childhood. *Acta Pathol Microbiol Scand c*. 1981;89:205–208.

76. Gregory JJ, Finlay JL. Alpha-fetoprotein and beta-human chorionic gonadotropin: their clinical significance as tumour markers. *Drugs*. 1999;57:463–467.

77. Cole LA, Sasaki Y, Muller CY. Normal production of human chorionic gonadotropin in menopause. *N Eng J Med*. 2007;356:1184. [Letter to the Editor].

78. De Backer B, Goffin F, Nisolle M, Minon JM. Persistent low hCG levels beyond pregnancy: report of two cases and review of literature. *Ann Biol Clin*. 2013;71:496–502. [article in French].

79. Cole LA. Phantom hcG and phantom choriocarcinoma. *Gynecol Oncol*. 1998;71:325–329.

80. Rotmensch S, Cole LA. False diagnosis and needless therapy of presumed malignant disease in women with false-positive human chorionic gonadotropin concentrations. *Lancet*. 2000;355(9205):712–725.

81. Axinte CI, Alexa T, Cracana I, Alexa ID. Macro-creatine kinase syndrome as an underdiagnosed cause of CK-MB increase in the absence of myocardial infarction: two case reports. *Rev Med-Chir Soc Med Nat Iasi*. 2012;116(4):1033–1038.

82. Sztefko K, Grodzicki T, Strach M. Macro-CK in 82 year old woman. A case report. *Int J Cardiol*. 2010;141. e57–578.

83. Michielsen EC, Bisschops PG, Janssen MJ. False positive troponin result caused by a true macrotroponin. *Clin Chem Lab Med*. 2011;49:923–925.

84. Warner JV, Marshall GA. High incidence of macrotroponin I with a high-sensitivity troponin I assay. *Clin Chem Lab Med*. 2016;54:1821–1829.

85. Herman DS, Ranjitkar P, Yamaguchi D, Grenache DG. Endogenous alkaline phosphatase interference in cardiac troponin I and other sensitive chemiluminescence immunoassays that use alkaline phosphatase activity for signal amplification. *Clin Biochem*. 2016;49:1118–1121.

86. Lum G, Solarz DE, Farney L. False positive cardiac troponin results in patients without acute myocardial infarction. *Lab Med*. 2006;37:546–550.

87. Shayanfar N, Bestmann L, Schulthess G, Hersberger M. False-positive cardiac troponin T due to assay interference with heterophilic antibodies. *Swiss Med Wkly*. 2008;138:470.

88. Janssen MJ, Velmans MH, Heesen WF. A patient with a high concentration of B-type natriuretic peptide (BNP) but normal N-terminal proBNP concentration: a case report. *Clin Biochem*. 2014;47:1136–1137.

89. Roy V. Artifactual laboratory abnormalities in patients with paraproteinemia. *South Med J*. 2009;102:167–170.

90. Smogorzewska A, Flood JG, Long WH, Dighe AS. Paraprotein interference in automated chemistry analyzers. *Clin Chem*. 2004;50. 1691–163.

91. Yang Y, Howanitz PJ, Howanitz JH, Gorfajn H, et al. Paraproteins are a common cause of interferences with automated chemistry methods. *Arch Pathol Lab Med*. 2008;132:217–223.

92. King RI, Florkowski CM. How paraproteins can affect laboratory assays: spurious results and biological effects. *Pathology*. 2010;42:397–401.

93. Jacobs JF, van der Molen RG, Bossuyt X, Damoiseaux J. Antigen excess in modern immunoassays: to anticipate on the unexpected. *Autoimmun Rev*. 2015;14:160–167.

94. Fernando SA, Wilson GS. Studies of the 'hook' effect in the one-step sandwich immunoassay. *J Immunol Methods*. 1992;151:47–66.

95. Dodig S. Interferences in quantitative immunochemical methods. *Biochem Med*. 2009;19:50–62.

96. Al-Mahdili HA, Jones GR. High-dose hook effect in six automated human chorionic gonadotrophin assays. *Ann Clin Biochem*. 2010;47:383–385.

97. Yadav YK, Fatima U, Dogra S, Kaushik A. Beware of "hook effect" giving false negative pregnancy test on point-of-care kits. *J Postgrad Med*. 2013;59:153–154.

98. Nodler JL, Kim KH, Alvarez RD. Abnormally low hCG in a complete hydatidiform molar pregnancy: the hook effect. *Gynecol Oncol Case Rep*. 2011;1:6–7.

99. Ranjitkar P, Turtle CJ, Harris NS, Holmes DT, et al. Susceptibility of commonly used ferritin assays to the classic hook effect. *Clin Chem Lab Med*. 2016;54:e41–e43.

100. Akamatsu S, Tsukazaki H, Inoue K, Nishio Y. Advanced prostate cancer with extremely low prostate-specific antigen value at diagnosis: an example of high dose hook effect. *Int J Urol*. 2006;13:1025–1027.

101. Butler SA, Cole LA. Use of heterophilic antibody blocking agent (HBT) in reducing false-positive hCG results. *Clin Chem*. 2001;47:1332–1333.

102. Reinsberg. Different Efficacy of various blocking reagents to eliminate interferences by HAMA in a 2-site immunoassay. *Clin Biochem*. 1996;29:145–148.

103. Bonetti A, Monica C, Bonaguri C, et al. Interference by heterophilic antibodies in immunoassay: wrong increase of myoglobin values. *Acta Biomed*. 2008;79:140–143.

104. Gessl A, Blueml S, Bieglmayer C, Marculescu R. Anti-ruthenium antibodies mimic macro-TSH in electro-chemiluminescent immunoassay. *Clin Chem Lab Med*. 2014;52:1589–1594.

CHAPTER 6

Issues of Interferences in Therapeutic Drug Monitoring

INTRODUCTION

Therapeutic drug monitoring is only required for drugs with a narrow therapeutic index. Therefore a relatively small number of drugs (approximately 50–60 drugs) require routine monitoring. The International Association of Therapeutic Drug Monitoring and Clinical Toxicology adopted the following definition for drug monitoring: "Therapeutic drug monitoring is defined as the measurement made in the laboratory of a parameter that, with appropriate interpretation, will directly influence prescribing procedures. Commonly, the measurement is in a biological matrix of a prescribed xenobiotic, but it may also be of an endogenous compound prescribed as a replacement therapy in an individual who is physiologically or pathologically deficient in that compound."[1] The characteristics of a drug where therapeutic drug monitoring is beneficial include

- difficulty in interpreting therapeutic or low toxicity of a drug based on clinical evidence alone,
- narrow therapeutic range where the dose of a drug that produces the desired therapeutic concentrations in one patient may cause toxicity in another patient,
- the drug is toxic but adverse drug reactions may be avoided by therapeutic drug monitoring,
- there is a good correlation between drug concentration and its therapeutic response or toxicity.

Traditionally therapeutic drug monitoring involves measuring drug concentration in a biological matrix most commonly serum or plasma and interpreting these concentrations in terms of relevant clinical parameters. Whole blood is the preferred matrix for therapeutic drug monitoring of immunosuppressants, except for mycophenolic acid. For success of therapeutic drug monitoring program, good communication among clinicians, laboratory professionals, and pharmacists is essential.[2] Therefore accurate determination of drug level in the proper biological matrix is essential for correct interpretation of test result and subsequent dose adjustment if needed. However, depending on the analytic method, significant interference may be encountered in therapeutic drug monitoring and such interference may falsely increase the true drug level (positive interference) in most cases but negative interference (falsely lower result) has also been reported. In general, negative interference is clinically more challenging because clinician may increase the dose without realizing that the drug concentration may be falsely lowered.

ANALYTIC METHODS OF THERAPEUTIC DRUG MONITORING

Various analytic methods such as immunoassay, gas chromatography with flame ionization of nitrogen detector (analysis of pentobarbital), gas chromatography combined with mass spectrometry, liquid chromatography (LC, high-performance liquid chromatography [HPLC]) combined with UV detector, LC combined with mass spectrometry (LC-MS), and LC combined with tandem MS (LC-MS/MS) are used for therapeutic drug monitoring.

Immunoassays are most commonly used for measuring concentrations of various drugs in serum, plasma, and less commonly in whole blood. Specimens can be analyzed in most cases directly without any pretreatment. Moreover, specimens can be batched and results can be obtained within 30–60 min from the beginning of the run. However, the antibody used in an immunoassay may cross-react with another molecule with a similar structure to the analyte drug molecule, most commonly the drug metabolite causing interference. In addition, other structurally related drugs and even endogenous compounds, such as high bilirubin concentrations, hemolysis, and elevated lipid levels, may also interfere with immunoassays. In contrast, chromatographic techniques are more labor intensive, requiring highly experienced medical technologists and high initial cost for acquiring such equipment, but such methods are relatively free from interferences. Therefore LC-MS or LC-MS/MS methods are used mostly for therapeutic monitoring of drugs where immunoassays are not commercially available, for example, monitoring of

Biotin and Other Interferences in Immunoassays. https://doi.org/10.1016/B978-0-12-816429-7.00006-X

newer antidepressants, antiretrovirals. Although immunoassays are commercially available for routine monitoring of cyclosporine, tacrolimus, sirolimus, everolimus (these immunosuppressants are monitored using whole blood), and mycophenolic acid (monitored in serum or plasma), certain large medical centers often use LC-MS/MS for monitoring of immunosuppressants due to significant interference of drug metabolites with assay-antibody causing falsely elevated drug levels.

INTERFERENCE FROM BILIRUBIN, HEMOLYSIS, AND LIPEMIA

Endogenous factors such as bilirubin, hemolysis, and high lipid levels if present in a specimen may potentially interfere with therapeutic drug monitoring using immunoassays. Usually total bilirubin concentration up to 20 mg/dL is not a problem for most assays used for therapeutic drug monitoring, but a bilirubin level over 20 mg/dL, which is uncommon, may cause some interference in certain immunoassays. High bilirubin concentrations may interfere with acetaminophen assay.

Case report: A severely jaundiced 17-year-old male presented to the emergency room with abdominal pain. He showed highly elevated serum bilirubin (19.8 mg/dL) but serology test results were negative for hepatitis A, B, and C. His serum acetaminophen concentration was 34 μg/mL indicating possibility of acetaminophen toxicity, although the patient denied taking any medication. Acetaminophen was measured by an enzymatic method (Roche Diagnostics) where hydrolysis of acetaminophen to *p*-aminophenol is catalyzed by arylacylamidase. Then *p*-aminophenol is condensed with *o*-cresol in the presence of periodate forming the blue indophenol chromophore, which is measured at 600 nm. The package insert indicated that bilirubin may interfere if present at a concentration exceeding 25 mg/dL. When authors analyzed 12 specimens collected from patients with high bilirubin (15.9–33.8 mg/dL) concentrations but not taking acetaminophen, apparent acetaminophen concentrations (5–18 μg/mL) were observed in these sample. Moreover, nonlinearity upon dilution confirmed assay interference. When bilirubin levels were reduced to <5 mg/dL, by dilution, acetaminophen concentrations also approached zero value.[3]

Polson et al.[4] also reported false-positive acetaminophen result in a patient with liver injury. Fong et al. reported persistently increased acetaminophen levels in a patient with acute liver failure. The authors used Vitros System (Ortho Clinical Diagnostics) for measurement of acetaminophen concentration in serum.[4]

In another study the authors reported that the degree of bilirubin interference in the acetaminophen assay using Beckman Coulter AU5822 analyzer is dependent on both bilirubin and acetaminophen concentrations. There was a decrease in the apparent acetaminophen concentration by an average 30% at a bilirubin concentration of 24.56 mg/dL (420 μmol/L).[5] Currently, the interference of elevated bilirubin levels in the colorimetric assay for salicylate (Trinder salicylate assay) is also problematic. However, immunoassay for salicylate is free from such interferences.[6]

Hemolysis can occur in vivo, during venipuncture and blood collection, or during processing of the specimen. Hemoglobin interference depends on its concentration in the specimen. Serum appears hemolyzed when the hemoglobin concentration exceeds 20 mg/dL. Hemoglobin interference is caused not only by the spectrophotometric properties of hemoglobin but also by its participation in chemical reaction with sample or reactant components as well. The absorbance maxima of the heme moiety in hemoglobin are at 540–580 nm. Methods that use the absorbance properties of NAD or NADH (340 nm) may thus be affected by hemolysis.

Lipids in serum or plasma exist as complexed with proteins called lipoproteins. Lipoproteins, consisting of various proportions of lipids, range from 10 to 1000 nm in size (the higher the percentage of the lipid, the lower the density and larger the particle size of the resulting lipoprotein). The lipoprotein particles with high lipid contents (such as chylomicrons and very-low-density lipoprotein) are micellar and are the main source of assay interference, especially in turbidimetric assays.

INTERFERENCES IN DIGOXIN IMMUNOASSAYS

In general, digoxin immunoassays suffer from more interferences than other immunoassays because serum digoxin level is very low (therapeutic range, 0.5–1.5 ng/mL) compared with other therapeutic drugs, for example, the therapeutic range of a commonly monitored drug phenytoin is 10–20 μg/mL. However, with the introduction of very specific monoclonal antibody in the assay design, many old issues of interference in digoxin immunoassays have been resolved. For example, in the 1980s and 1990s, significant interferences by endogenous digoxin like immunoreactive substances (DLISs) in serum digoxin measurement have been reported. DLIS levels are relatively low in healthy people, causing no assay interference but they are significantly elevated in patients with volume expansion, such as those with uremia, liver disease, and hypoalbuminemia; pregnant women; and premature babies. However, one of the assays most affected by DLIS, the fluorescence polarization immunoassay digoxin assay marketed by Abbott Laboratories (Abbott Park, IL), has been discontinued. In addition, newer digoxin assays are free from such interference, for example, iDigoxinchemiluminescent microparticle immunoassay (CMIA) for digoxin manufactured by the Abbott Laboratories for application on the Architect analyzer is virtually free from DLIS interference. Lampon et al. measured apparent digoxin concentrations using iDigoxin assay and Architect analyzer in digoxin-free pregnant women ($n = 50$); patients with renal insufficiency ($n = 50$) and liver disease ($n = 50$); kidney transplant ($n = 25$), liver transplant ($n = 50$), and critically ill patients ($n = 50$) and observed no apparent digoxin concentration (values below the level of quantitation of 0.30 ng/mL) in any of them. These patients were expected to have high DLIS levels. The authors concluded that the iDigoxin assay manufactured by the Abbott Laboratories would be relatively free from endogenous DLIS interference.[7]

Although various diuretics interact with digoxin and may increase serum digoxin level, potassium-sparing diuretics, such as spironolactone, potassium canrenoate, and eplerenone, not only pharmacokinetically interact with digoxin but also may interfere with serum digoxin measurements producing falsely elevated digoxin level using various digoxin immunoassays, as well as falsely lower digoxin values using the microparticle enzyme immunoassay (MEIA) II digoxin assay (Abbott Laboratories).[8] Monitoring free digoxin may eliminate some interference. However, relatively new digoxin assays utilizing specific monoclonal antibodies against digoxin are free from such interferences.[9] DeFrance et al. reported that digoxin assays for application on Architect clinical chemistry platforms (cDig, particle-enhanced turbidimetric inhibition immunoassay [PETINIA]) and Architect immunoassay platforms (iDig, CMIA), both from Abbott Diagnostics, are free from interferences by spironolactone, potassium canrenoate, and their common metabolite canrenone.[10] In another study, the authors reported statistically significant but clinically insignificant interference of spironolactone and related compounds with the luminescent oxygen channeling assay (LOCI) for digoxin for application on the Vista 1500 analyzer (Siemens).[11]

The major metabolites of digoxin are digoxigenin, digoxigenin monodigitoxoside, digoxigenin bisdigitoxoside, and dihydrodigoxin. These metabolites exhibit significantly different cross-reactivities against various antidigoxin antibodies, but due to their relatively low concentrations in serum, they may not have any significant effect on digoxin immunoassays. However, overestimation of serum digoxin concentrations in patients with renal disease, liver disease, or diabetes has been reported probably due to elevated levels of digoxin metabolites in serum.[12]

The presence of human anti-animal antibody (heterophilic antibody), especially those directed against mouse, in serum may cause interference with certain immunoassays. The clinical use of mouse monoclonal antibody for radioimaging and treatment of certain cancers may cause accumulation of human antimouse antibody. Anti-animal antibodies are also found among veterinarians, farm workers, or pet owners due to exposure to animals, and these antibodies are broadly classified as heterophilic antibodies. Usually the presence of heterophilic antibodies in serum may interfere with sandwich assays designed for measuring relatively large molecules, such as β-human chorionic gonadotropin. Nevertheless, Liendo et al. described a case report of a patient with cirrhotic liver disease and atrial fibrillation being treated with spironolactone and digoxin, who showed a toxic digoxin concentration of 4.2 ng/mL without any symptom of digoxin toxicity. After discontinuation of both drugs, serum digoxin level stayed elevated (over 3.0 ng/mL) for approximately 5 weeks. The authors were able to remove this interference by measuring free digoxin and speculated that falsely elevated digoxin levels in the patient were due to interference of heterophilic antibodies.[13] In another report the authors observed a moderate positive interference of heterophilic antibody (0.8 ng/mL apparent digoxin) on the Dimension DGNA digoxin immunoassay (captures rabbit antibody; marketed by DadeBehring, now Siemens) in an elderly patient with a history of social cat handling. The interference was eliminated by pretreatment of specimen with a heterophilic antibody–blocking reagent. The authors also reported that the digoxin immunoassay for application on the Architect analyzer (Abbott Laboratories) was not affected by heterophilic antibodies.[14]

Digibind/DigiFab Interference With Digoxin Immunoassays

Digibind and DigiFab are Fab fragments of antidigoxin antibody used in treating life-threatening acute digoxin overdose. Digibind was marketed in 1986, while DigiFab was approved for use in 2001. The molecular weight of DigiFab (46,000 Da) is similar to the molecular weight of Digibind (46,200 Da), and both compounds can be excreted in urine. Both

Digibind and DigiFab interfere with serum digoxin measurement using immunoassays. The magnitude of interference depends on the assay design and the specificity of the antidigoxin antibody. McMillin et al. studied the effect of Digibind and DigiFab on 13 different digoxin immunoassays. The magnitude of interference varied significantly with each method. The authors commented that monitoring free digoxin (in the protein-free ultrafiltrate) eliminated this interference because both Digibind and DigiFab, due to their higher molecular weight, are absent in the protein-free ultrafiltrate.[15]

Case Report

A 35-year-old woman intentionally swallowed 100 Lanitop tablets (0.1 mg methyldigoxin per tablet) in a suicide attempt. Methyldigoxin is a semisynthetic cardiac glycoside that is rapidly converted into digoxin after oral administration. On admission, approximately 19 h after ingestion, her serum digoxin level was 7.4 ng/mL and the patient was treated immediately with 80 mg of Digibind. A total of 395 mg of Digibind was administered to the patient. Her total serum digoxin level peaked at 125 ng/mL 23 h after ingestion, but the free serum digoxin level immediately decreased to nontoxic levels, indicating that Digibind therapy was effective in treating her overdose. Her symptoms of digoxin toxicity (electrocardiogram) as well as nausea and vomiting resolved within 3 h of initiation of therapy, and the patient was discharged from the hospital after 3 days.[16]

Interference of Herbal Supplements With Digoxin Immunoassays

In the United States, complementary and alternative medicines are classified as dietary supplements and marketed pursuant to the Dietary Supplement Health and Education Act of 1994. Complementary and alternative medicines, including Ayurvedic medicines, are becoming increasingly popular in the United States, Europe, and other parts of the world. Significant interferences of Chinese medicines Chan Su, Lu-Shen-Wan, oleander-containing herbal products, and supplements containing lily of the valley with various digoxin assays have been reported.

Chan Su, is prepared from the dried white secretion of the auricular glands and the skin glands of Chinese toads (*Bufo melanostictus* Schneider or *Bufo gargarizans* Gantor) and used for treating various heart diseases in traditional Chinese medicine. Bufalin, the active component of Chan Su, is responsible for interference with various digoxin assays because of structural similarity with digoxin. The interference of Chan Su and Lu-Shen-Wan in serum digoxin measurement can be positive (falsely elevated digoxin concentrations) or negative (falsely lower digoxin concentration) depending on the assay design. Although Beckman assay (SYNCHRON LX system, Beckman Coulter) and Roche assay (Tina-quant) showed positive interference in the presence of Chan Su, the MEIA digoxin assay on the AxSYM platform (Abbott Laboratories) showed negative interference of Chan Su in serum digoxin measurement. However, the components of Chan Su responsible for digoxin like immunoreactivity are strongly bound to serum proteins (>90%) and are virtually absent in the protein-free ultrafiltrate. Therefore measuring free digoxin concentration in the protein-free ultrafiltrate could be used to mostly eliminate the interference of Chan Su and Lu-Shen-Wan in serum digoxin measurements.[17]

The oleanders are evergreen ornamental shrubs that grow in the Southern parts of the United States from Florida to California, Australia, India, Sri Lanka, China, and other parts of the world. All parts of both types of oleander plants are toxic. Boiling or drying the plant does not inactivate the toxins. Many cardiac glycosides are present in the oleander plant, but oleandrin is mostly responsible for interference with digoxin immunoassays.

Case Report

A 30-year-old woman complained of weakness, nausea, and vomiting after drinking a herbal tea she believed to be prepared from eucalyptus leaves. On arrival, paramedics found that the patient had slurred speech and complained of numbness in her tongue. Her husband described her confused. In the emergency room the patient was orotracheally intubated. Despite lifesaving attempts, she later died. The toxicologic study during postmortem examination showed serum digoxin concentration of 6.4 ng/mL, although she was not on digoxin and did not receive any digoxin during her treatment in the hospital prior to her death. After further investigation the cause of death was reported to be due to oleander poisoning.[18]

Taking advantage of the cross-reactivity of oleandrin, the active component of oleander with structural similarity to digoxin, oleander poisoning may be detected indirectly by observing apparent digoxin concentration in the serum of a patient suspected with oleander poisoning but not taking digoxin.[19] Bidirectional (negative interference with lower oleandrin concentration but positive interference with higher oleandrin concentration) interference with the LOCI digoxin assay has also been reported.[20]

The lily family is composed of 280–300 genera made up of 4000–4600 different species but only 90 genera representing approximately 525 species are found in North America. Lilies are popular decorative plants and

TABLE 6.1
Interferences in Digoxin Immunoassays

Interfering Compound	Magnitude of Interference	Comments
Spironolactone and its metabolite canrenone	Modest	Newer assays are virtually free from such interferences.
Heterophilic antibody	Significant	There are only a few reports of such interference. However, monitoring free digoxin eliminates such interference.
Digibind/DigiFab	Significant	This interference can be eliminated by monitoring free digoxin level.
Chan Su, Lu-Shen-Wan (Chinese medicine used as a tonic to the heart)	Significant	Interference is due to active compound bufalin, which has structural similarity with digoxin.
Oleander (oleander extract is used in herbal supplements as a heart stimulant)	Significant	Oleandrin, the active component of oleander extract, interferes with digoxin immunoassays due to structural similarity with digoxin
Lily of the valley extract (extract is used in herbal supplements as a heart stimulant)	Significant	Convallatoxin, the active component of lily of the valley, has structural similarity with digoxin.

are also found in floral arrangements. The *Convallaria* genus, commonly called "lily of the valley," is used for decoration in the United States. The entire plant is toxic, containing cardiac glycosides, causing digitalis-like toxicity. The principle toxic cardiac glycoside found in lily of the valley is convallatoxin. There are several reports of lily of the valley poisoning.[21] Symptoms of digitalis-like toxicity in a family after accidental ingestion of lily of the valley plant have also been reported.[22] Despite toxicity, lily of the valley extract is used in herbal medicine as a tonic for heart.[23] Convallatoxin has structural similarity with digoxin and interferes with the LOCI digoxin assay (Siemens).[24] Interestingly, the iDigoxin assay on the Architect analyzer (Abbott Laboratories) is more sensitive than the LOCI digoxin assay in detecting the presence of convallatoxin.[25] Fink et al.[26] also compared five digoxin assays for rapid detection of convallatoxin. However, any digoxin immunoassay capable of rapid detection of convallatoxin is also unsuitable for therapeutic drug monitoring of digoxin in patients who are taking lily of the valley herbal supplements. Various compounds known to interfere with digoxin immunoassays are summarized in Table 6.1.

propranolol are also subjected to therapeutic drug monitoring. Digoxin-quinidine interaction is also an important consideration for therapeutic drug monitoring of both quinidine and digoxin. An increase in serum digoxin concentration occurs in 90% patients receiving quinidine. On average, serum digoxin concentration doubles during treatment with therapeutic doses of quinidine because quinidine impairs renal clearance of digoxin and may also decrease its volume of distribution. Therefore dose adjustments based on therapeutic drug monitoring of both digoxin and quinidine are needed to avoid adverse drug reactions.[27] Amiodarone contains iodine and thyroid disorder may occur in some patients being treated with amiodarone for a prolonged period (although most patients are euthyroid). Therefore monitoring of thyroid function is recommended for patients taking this medication. Procainamide is metabolized to N-acetylprocainamide (NAPA), an active metabolite. Therefore monitoring of both procainamide and NAPA is recommended. Less frequently monitored cardioactive drugs are analyzed using HPLC or LC-MS/MS and such methods are virtually free from interferences.

ISSUES OF INTERFERENCES WITH MONITORING OF OTHER CARDIOACTIVE DRUGS
In addition to digoxin, other cardioactive drugs such as procainamide, quinidine, disopyramide, lidocaine, mexiletine, flecainide, tocainide, propafenone, and

ISSUES OF INTERFERENCE WITH MONITORING OF ANTIEPILEPTIC DRUGS
Monitoring classical anticonvulsants, such as phenytoin, carbamazepine, phenobarbital, and valproic acid, is essential for proper patient management, and immunoassays are commercially available for determination

of serum or plasma concentrations of all these drugs. Usually phenobarbital and valproic acid immunoassays are robust, and interference is observed only rarely. Overdose with amobarbital or secobarbital may cause a falsely elevated phenobarbital level, but it is observed rarely. However, carbamazepine and phenytoin immunoassays are subjected to interferences.

Carbamazepine is metabolized into carbamazepine-10,11-epoxide, which is an active metabolite. This metabolite cross-reacts with carbamazepine immunoassay and cross-reactivity may be as high as 85%–90% in the PETINIA (Siemens) carbamazepine assay but the enzyme multiplied immunoassay technique (EMIT) is virtually free from this interference. Increased concentration of carbamazepine-10,11-epoxide is observed in patients with renal insufficiency as well as in patients overdosed with carbamazepine. Falsely elevated carbamazepine level measured using the PETINIA may be observed in patients with elevated carbamazepine-10,11-epoxide levels.

Case Report

A 49-year-old man with a history of bipolar disorder arrived unconscious at the emergency room after taking an unknown amount of Tegretol 200-mg tablet. The serum carbamazepine level measured by the PETINIA was 45.1 μg/mL. When the sample was retested using HPLC and EMIT the values were 33.6 and 27.8 μg/mL, respectively. The carbamazepine-10,11-epoxide concentration measured by HPLC was 12.5 μg/mL. Gastric lavage was followed by activated charcoal decontamination and the status of the patient was improved. After 12 h, his serum carbamazepine concentration was 29.5 μg/mL as measured by EMIT but 26.2 μg/mL as measured by HPLC. In contrast, PETINIA failed to show significant reduction in carbamazepine concentration following treatment, as the serum carbamazepine value was still elevated at a level of 42.5 μg/mL due to elevated carbamazepine-10,11-epoxide concentration of 18.2 μg/mL.[28]

Hydroxyzine and cetirizine are antihistamine drugs that interfere only with the PETINIA carbamazepine assay. Parant et al. reported falsely toxic serum carbamazepine concentration of 40.6 μg/mL (therapeutic range, 4–12 μg/mL) in a 32-year-old male using PETINIA. However, a subsequent drugscreening analysis by LC-diode array detection showed very low carbamazepine and epoxide levels (<0.5 μg/mL) but the patient had toxic hydroxyzine concentration (0.55 μg/mL; therapeutic range, <0.1 μg/mL). No other drugs or alcohol were detected. On questioning, the patient admitted regularly taking Atarax (hydroxyzine) to feel

"high." A diagnosis of hydroxyzine toxicity was made, and it was determined that the false-positive carbamazepine level measured by PETINIA was due to interference of hydroxyzine. However, EMIT was free of such interference.[29] In addition, carbamazepine assay on the ADVIA Centaur analyzer (Siemens) is also free from hydroxyzine and cetirizine interference.[30]

Fosphenytoin, a prodrug of phenytoin is rapidly converted into phenytoin after administration. Fosphenytoin, unlike phenytoin, is readily water soluble and can be administered intravenously or intramuscularly. Unlike phenytoin, fosphenytoin does not crystallize at the injection site and no discomfort is experienced by the patient. Fosphenytoin is not typically monitored clinically because of its short half-life and lack of pharmacologic activity. However, phenytoin is monitored in a patient after administration of fosphenytoin, but in this case, monitoring of phenytoin must be initiated after complete conversion of fosphenytoin into phenytoin. In uremic patients, after fosphenytoin administration, phenytoin concentrations measured by various immunoassays may be significantly higher than the true phenytoin level determined by HPLC. Annesley et al.[31] identified a unique oxymethylglucuronide metabolite derived from fosphenytoin in sera of uremic patients and demonstrated that this unusual metabolite was responsible for the cross-reactivity with various phenytoin immunoassays.

Paraproteins, especially IgM, if present in high concentrations may cause falsely lower serum phenytoin value when measured by PETINIA. A 69-year-old man with alcoholic liver disease was admitted to the authors' hospital for uncontrolled seizure. The patient was treated with fosphenytoin to control seizure but despite therapy, his serum phenytoin levels were <1 μg/mL using PETINIA. However, serum phenytoin levels were between 5 and 10 μg/mL when measured by HPLC. Plasma levels of IgM and IgG measured by ELISA, showed normal IgG level, but elevated IgM level (2–3 times above the upper limit of normal). The authors concluded that the elevated IgM level was responsible for negative interference in phenytoin measurement using PETINIA.[32] Interferences in therapeutic drug monitoring of carbamazepine and phenytoin using immunoassays are listed in Table 6.2.

ISSUES OF INTERFERENCE IN MONITORING OF SELECTED ANTIBIOTICS

In general, aminoglycosides and vancomycin are monitored routinely using appropriate immunoassays. These immunoassays are robust and virtually free from

TABLE 6.2
Interferences in Therapeutic Drug Monitoring of Carbamazepine and Phenytoin Using Immunoassays

Anticonvulsant	Interfering Compound	Comments
Carbamazepine	Carbamazepine-10,11-epoxide	Falsely elevated serum carbamazepine concentration in overdosed patients due to interference of this carbamazepine metabolite with assay antibody in the PETINIA (Siemens) assay. However, EMIT is not affected.
Carbamazepine	Hydroxyzine and cetirizine	These drugs interfere at high concentrations expected only in overdosed patients with PETINIA (Siemens) assay only.
Phenytoin	Fosphenytoin	Fosphenytoin is a prodrug of phenytoin. Fosphenytoin administration falsely elevates serum phenytoin levels in immunoassays due to formation of oxymethylglucuronide metabolite usually present only in uremic patients.
Phenytoin	Paraprotein (IgM)	High concentrations of paraproteins, especially IgM, may falsely lower serum phenytoin levels measured by PETINIA.
Valproic acid	Paraprotein (IgM)	May cause interference in valproic acid immunoassay for application on Beckman analyzer due to high blank reading.

EMIT, enzyme multiplied immunoassay technique; *PETINIA*, particle-enhanced turbidimetric inhibition immunoassay.

interferences. Gentamicin is not a single molecule but a complex of three major (C1, C1a, and C2) and several minor components. In addition, the C2 component is a mixture of stereoisomers. Most immunoassay methods can measure the total gentamicin concentration in serum or plasma but are incapable of measuring individual components. There are only a few case reports of interference of paraproteins in measuring vancomycin and gentamicin serum levels using specific immunoassays. The interference could be negative or positive depending on the immunoassay used.

Case Report

A 68-year-old woman with a history of lymphoplasmacytic lymphoma with an IgM kappa monoclonal component of 42.8 g/L was started on vancomycin, and on day 3, her trough vancomycin concentration was <0.1 µg/mL, as measured by Beckman Coulter SYNCHRON competitive turbidimetric immunoassay, which was inconsistent with vancomycin therapy. When the specimen was sent to another laboratory and analyzed using a competitive enzyme-linked immunoassay (Olympus analyzer) a value of 9.8 µg/mL was obtained, indicating that vancomycin level was falsely lowered (negative interference) due to the presence of paraproteins in the specimen.[33]

Case Report

An 82-year-old man diagnosed with Waldenström disease a year ago was admitted to the hospital with anorexia, asthenia, and hyperviscosity syndrome. Hemocultures showed methicillin-resistant *Staphylococcus aureus* sepsis and he was treated with vancomycin. On the first day of vancomycin therapy, his serum vancomycin level was >100 µg/mL (measured by iVancomycin assay on the Architect analyzer, Abbott Laboratories), which was inconsistent with the vancomycin dose. Surprisingly, sample dilution showed nonlinearity, an indication of assay interference. In addition, measuring vancomycin using 1:4 and 1:5 dilution showed a much lower calculated serum vancomycin level of 21.8 and 17.2 µg/mL, respectively. Analysis of specimen collected from this patient prior to vancomycin therapy showed a false vancomycin level of 70.3 µg/mL using iVancomycin assay, thus further confirming assay interference. However, when the same specimen obtained prior to vancomycin therapy was analyzed using cobas c502 analyzer (Roche) and LC-MS/MS, values were nondetected as expected. Moreover, reanalyzing specimen from day 1 using LC-MS/MS (19.5 µg/mL) and cobas c502 analyzer (18.6 µg/mL) showed expected concentration, thus further verifying falsely elevated vancomycin level measured by iVancomycin assay. The authors speculated that this positive interference in vancomycin measurement was due to elevated IgM level of 7.4 g/dL. However, this interference was eliminated by monitoring free vancomycin because IgM, owing to its high molecular weight, was absent in the protein-free ultrafiltrate.[34]

LeGatt et al. studied the effect of various immunoglobulins on vancomycin measurement using four immunoassays and observed that IgA and IgG levels in serum and plasma did not affect any of the vancomycin immunoassays. Although IgM did not affect the fluorescence polarization immunoassay and enzyme multiplied immunoassay methods, it did attenuate vancomycin concentrations by both the PETINIA (Siemens and Beckman Coulter) assays, with a more pronounced effect on the latter, producing concentrations >20% lower than expected in the patient serum and spiked plasma pools. The effect was progressively negative at effective IgM concentrations of 10 and 15 mg/L.[35] In another study the authors reported falsely lower vancomycin levels in a 64-year-old woman with a history of multiple immune-related comorbidities, who received vancomycin for treatment of a prosthetic joint infection growing coagulase-negative *Staphylococcus* spp., and a 33-year-old man with a history of Felty syndrome, who received vancomycin for the treatment of methicillin-resistant *Staphylococcus aureus* pneumonia. Both patients had multiple vancomycin trough concentrations determined using the Beckman Coulter PETINIA method and had measured concentrations reported as less than 4 mg/L, despite appropriate vancomycin dosing for their body weights, age, and organ function. When sera collected from these patients were tested by alternative methods, expected vancomycin level consistent with dosing was observed, thus confirming negative interference of paraproteins in serum vancomycin measurement using PETINIA.[36] IgM paraprotein was also identified as the cause of interference with the gentamicin, vancomycin, and valproate assays using Beckman DxC800 and DC800 analyzers.[37]

ISSUES OF INTERFERENCE IN MONITORING OF TRICYCLIC ANTIDEPRESSANTS

The major interference in immunoassays for determining tricyclic antidepressant (TCA) concentrations in serum or plasma is from the TCA metabolites. For example, nortriptyline, a metabolite of amitriptyline, has a very high cross-reactivity (almost 100%) with TCA immunoassay similar to the parent drug. Therefore in a patient taking amitriptyline, TCA immunoassay would provide the total TCA level (amitriptyline + nortriptyline). Therefore for therapeutic drug monitoring of TCA, chromatographic methods must be used. TCA immunoassays are only useful in diagnosis of TCA overdose where the combined value of drug and metabolite indicates TCA overdose.

Several drugs are known to interfere with TCA immunoassays. Phenothiazine and its metabolites may interfere with TCA immunoassays, and even a therapeutic concentration of such drug may cause falsely elevated serum TCA level.[38] Carbamazepine is metabolized to carbamazepine-10,11-epoxide, an active metabolite. Both the parent drug and the epoxide metabolite interfere with TCA immunoassay because of structural similarities. Another structurally related drug, oxcarbazepine, also interferes with TCA immunoassays.[39,40] Cyclobenzaprine, a muscle relaxant, is structurally similar to TCAs and is also known to interfere with TCA immunoassays.[41] Results from in vitro study indicate that high serum levels of hydroxyzine and cetirizine may cause false-positive test results, with the fluorescence polarization immunoassay for TCA.[42] Quetiapine may also interfere with TCA immunoassays.[43] However, interference of diphenhydramine (Benadryl) and cyproheptadine with TCA immunoassays occurs only in overdosed patients.[44] Issues of interference in measurement of vancomycin, gentamicin, and TCA immunoassays are summarized in Table 6.3.

METABOLITE CROSS-REACTIVITY: A MAJOR PROBLEM WITH IMMUNOASSAYS FOR IMMUNOSUPPRESSANT DRUGS

Immunosuppressant drugs such as cyclosporine, tacrolimus, sirolimus, and everolimus must be monitored in whole blood, while mycophenolic acid is the only immunosuppressant that is monitored in serum or plasma. Chromatographic methods, especially LC-MS or LC-MS/MS, are the gold standard for therapeutic drug monitoring of immunosuppressants because immunoassays usually show significant positive bias compared with LC-MS/MS values due to significant metabolite cross-reactivities.

Most immunoassays for cyclosporine, tacrolimus, sirolimus, and everolimus require specimen pretreatment to extract drug from the whole blood, except antibody-conjugated magnetic immunoassays (ACMIA, Siemens) for cyclosporine, tacrolimus, and sirolimus, which include online mixing and ultrasonic lysis of whole blood. However, immunoassays for mycophenolic acid do not require serum pretreatment because the assay can be run using serum or plasma. Usually acyl glucuronide (a minor active metabolite of mycophenolic acid) is responsible for interferences in some immunoassays.

Interferences in Cyclosporine and Tacrolimus Immunoassays

Positive bias up to 40% was reported in older cyclosporine immunoassays when values were compared with

TABLE 6.3
Issues of Interference in Measurement of Vancomycin, Gentamicin, and Tricyclic Antidepressants

Drug	Interfering Substances	Comments
Vancomycin	Paraproteins	Paraprotein (elevated IgM) may cause both positive (falsely elevated value using iVancomycin assay) and negative interferences (both PETINIA assay; Beckman and Siemens) depending on the assay type.
Gentamicin	Paraprotein	Paraprotein (IgM) interferes with Beckman Coulter assay for gentamicin using DxC800 and DC800 analyzers by producing high blank readings.
TCA	Phenothiazines and metabolites, carbamazepine and its active metabolite carbamazepine-10, 11-epoxide, oxcarbazepine, cyclobenzaprine, quetiapine, cyproheptadine, diphenhydramine, hydroxyzine, cetirizine	Cyproheptadine, diphenhydramine (Benadryl), hydroxyzine, and cetirizine interfere only if present in serum representing toxic levels.

levels obtained by chromatography-based method. However, newer cyclosporine immunoassays are less prone to cross-reactivities from cyclosporine metabolites. Wallemacq et al. reported findings from multicenter evaluation of Abbott Architect cyclosporine CMIA using seven clinical laboratories. Values obtained by the immunoassay were compared with corresponding values obtained by LC-MS/MS. The authors observed minimal cross-reactivity of cyclosporine metabolites in the cyclosporine CMIA, as the two major cyclosporine metabolites AM1 and AM9 exhibited −2.5% to 0.2% and 0.8%–2.2% cross-reactivity, respectively, with the immunoassay. Comparison testing with Roche Integra showed 2.4% cross-reactivity for AM1 metabolite and 10.7% cross-reactivity for the AM9 metabolite. The authors concluded that the cross-reactivity of cyclosporine metabolites has been significantly reduced in the cyclosporine CMIA.[45] Soldin et al. evaluated the performance of a new ADVIA Centaur cyclosporine immunoassay that requires a single-step extraction and observed excellent correlation between cyclosporine values obtained by LC-MS/MS and ADVIA Centaur cyclosporine assay. The authors concluded that the new ADVIA Centaur assay compared favorably with LC-MS/MS.[46] However, falsely elevated blood cyclosporine levels due to presence of endogenous antibody have been reported with the ACMIA cyclosporine assay run on the Dimension RXL analyzer (Siemens). De Jonge et al. reported a falsely elevated cyclosporine level of 492 ng/mL in a 77-year-old patient. However, using LC-MS, the cyclosporine level was undetectable. In addition, Architect cyclosporine assay also yielded a value lower than the detection limit. Treating the specimen with polyethylene glycol and remeasuring cyclosporine levels in the

supernatant by the same ACMIA showed none detected level of cyclosporine, confirming that the interfering substance was a protein, most likely an endogenous antibody such as heterophilic antibody.[47]

In one study the author observed an average 18% negative bias when cyclosporine values obtained by LC-MS/MS were compared with values obtained by the cyclosporine CMIA. For tacrolimus, average values obtained by LC-MS/MS were 14% lower than the values obtained by the CMIA tacrolimus assay.[48] Leung et al.[49] reported an average 17% positive bias with Quantitative Microparticle System (QMS) tacrolimus assay (Thermo Fisher) compared with values obtained by LC-MS/MS. However, older tacrolimus immunoassays showed much higher bias when values obtained by immunoassays were compared with LC-MS/MS values. In addition, false-positive tacrolimus concentrations in patients with low hematocrit values and high imprecision at tacrolimus values less than 9 ng/mL with the MEIA tacrolimus assay for application on the AxSYM platform (Abbott Laboratories) have also been reported. However, EMIT was not affected.[50] As expected, older immunoassays for tacrolimus were also affected by the cross-reactivity from tacrolimus metabolites. Westely et al. evaluated CEDIA tacrolimus assay by measuring values obtained by CEDIA, LC-MS/MS, and MEIA. The authors observed a 33.1% bias with CEDIA and 20.1% bias with MEIA in tacrolimus values compared with the tacrolimus values measured by LC-MS/MS in renal transplant recipients.[51]

Like cyclosporine, the ACMIA tacrolimus assay is also affected by rheumatoid factors and endogenous heterophilic antibodies.[52] Altinier et al. described the interference of heterophilic antibody in the ACMIA tacrolimus assay. Samples from a patient showed tacrolimus values in the range of 49–12.5 ng/mL even after discontinuation of tacrolimus therapy. The authors confirmed that the elevated tacrolimus levels were due to the presence of heterophilic antibody, and by treating the samples with heterophilic blocking tubes and protein G resin, the interference was removed.[53] In another report the authors observed a high tacrolimus value (79.7 ng/mL) in a liver transplant recipient using ACMIA tacrolimus assay despite discontinuation of tacrolimus therapy. The authors identified β-galactosidase antibodies as the cause of interference because in this assay anti-tacrolimus antibody conjugated to β-galactosidase is used.[54] Rostaing et al. observed falsely elevated tacrolimus level of 24 ng/mL using the ACMIA tacrolimus assay, but none detected the level of tacrolimus using both LC-MS/MS and EMIT tacrolimus assay. The authors identified positive anti–double-stranded DNA autoantibodies as the cause of interference in ACMIA.[55]

Case Report

A 59-year-old man underwent a kidney transplant and was managed with tacrolimus and corticosteroids. For the first 3 weeks after transplant the patient's tacrolimus whole blood concentrations were consistent with the dose and were below 12 ng/mL. Then 25 days after transplant, his tacrolimus level measured by the ACMIA tacrolimus assay was found to be highly elevated to 21.5 ng/mL. Tacrolimus was discontinued but its level was still elevated. Suspecting interference, tacrolimus was analyzed using MEIA (Abbott Laboratories) and the observed value was below 2 ng/mL, indicating interference in tacrolimus measurement using ACMIA. The authors suggested that if the tacrolimus value measured by ACMIA does not match the clinical picture, tacrolimus must be measured by an alternative method before any clinical intervention, and the interference was probably due to the presence of heterophilic antibodies.[56]

Examples of metabolite interferences in some immunosuppressant monitoring using immunoassays are listed in Table 6.4.

TABLE 6.4
Examples of Metabolite Interferences in Some Immunosuppressant Monitoring Using Immunoassays

Immunosuppressant Immunoassays	Interfering Substances	Comments
Older cyclosporine assays (CEDIA, EMIT, etc.)	Cyclosporine metabolites	Average 17% positive bias compared with chromatography-based methods.
Newer cyclosporine assays (ADVIA Centaur, CMIA, etc.)	Cyclosporine metabolite	Significantly less bias compared with older assays.
ACMIA cyclosporine	Endogenous antibody/ cyclosporine metabolites	This is the only cyclosporine assay where values are falsely elevated due to the presence of endogenous antibodies such as heterophilic antibodies. ACMIA also shows cross-reactivity with cyclosporine metabolites.
Older tacrolimus assays (MEIA)	Low hematocrit Tacrolimus metabolite	False-positive tacrolimus result when haematocrit <25%. Also average 20.1% positive bias due to metabolite cross-reactivity.
CEDIA tacrolimus	Tacrolimus metabolites	Average 33.1% positive bias compared with chromatographic methods.
Newer tacrolimus assays (CMIA tacrolimus, etc.)	Tacrolimus metabolites	Average 20% positive bias compared with chromatographic methods.
ACMIA tacrolimus	Endogenous antibody/ tacrolimus metabolites	This is the only tacrolimus assay where values are falsely elevated due to the presence of endogenous antibodies such as heterophilic antibodies. ACMIA also shows cross-reactivity with cyclosporine metabolites.
QMS tacrolimus assay	Tacrolimus metabolites	Average 17% positive bias compared with chromatographic methods.
CMIA tacrolimus	Tacrolimus metabolites	Average 20% positive bias compared with chromatographic methods.

TABLE 6.4
Examples of Metabolite Interferences in Some ImmunosuppressantMonitoring Using Immunoassays—cont'd

Immunosuppressant Immunoassays	Interfering Substances	Comments
CMIA sirolimus	Sirolimus metabolites	Lowest reported positive bias was 14%. However, this assay has a high cross-reactivity with structurally similar everolimus.
QMS everolimus assay	Everolimus metabolites	Average 11% positive bias compared with chromatographic methods. However, this assay has an average 46% cross-reactivity with structurally similar sirolimus.
CEDIA mycophenolic acid	Mycophenolic acid metabolite	Significant cross-reactivity with mycophenolic acid glucuronide causing average positive bias of 18%.
PETINIA on Dimension EXL analyser	Mycophenolic acid metabolite	Average positive bias of 12% compared with an HPLC-UV method.
EMIT 2000 mycophenolic acid	Mycophenolic acid metabolite	No cross-reactivity reported with mycophenolic acid metabolite, values compared well with those from the chromatographic method.
Roche total mycophenolic acid	Mycophenolic acid metabolite	No cross-reactivity reported with mycophenolic acid metabolite, values compared well with those of the chromatographic method.

ACMIA, antibody-conjugated magnetic immunoassay; *CMIA*, chemiluminescent microparticle immunoassay; *EMIT*, enzyme multiplied immunoassay technique; *HPLC-UV*, high-performance liquid chromatographycombined with ultraviolet detection; *MEIA*, microparticle enzyme immunoassay; *PETINIA*, particle-enhanced turbidimetric inhibition immunoassay; *QMS*, Quantitative Microparticle System.

Interferences in Sirolimus and Everolimus Immunoassays

Schmidt et al. evaluated the new CMIA sirolimus assay for application on the Architect analyzer (Abbott Laboratories) and concluded that the assay only cross-reacts with sirolimus metabolite F4 and F5 but hematocrit has no effect on the assay. In a multisite clinical trail the authors observed an average of 14%, 25%, and 39% mean bias with CMIA and three different LC-MS/MS methods for the determination of sirolimus in three different sites that evaluated CMIA. The authors concluded that although CMIA correlated well with LC-MS/MS method, it shows a positive bias in sirolimus values compared with the values determined by more specific LC-MS/MS assays.[57]

QMS everolimus assay received FDA approval for clinical use in 2011 and the assay is linear between 1.5 and 20 ng/mL, covering the entire therapeutic range of everolimus. The limit of quantitation for this assay is 1.3 ng/mL. According to one report, this assay is not affected by commonly used 70 drugs, but structurally similar sirolimus exhibited an average 46% cross-reactivity. In addition, the average bias everolimus values determined by the QMS everolimus assay and the corresponding values obtained by a specific LC-MS/MS method were 11%

based on the comparison of 90 specimens obtained from patients receiving everolimus. The authors concluded that QMS everolimus assay showed adequate sensitivity and specificity and can be used for routine therapeutic drug monitoring of everolimus.[58]

Issues in Interference in Mycophenolic Acid Immunoassays

Mycophenolic acid is monitored in serum or plasma. Dasgupta et al. reported average positive bias of 12% with PETINIA mycophenolic acid on Dimension EXL analyzer in comparison to the HPLC combined with ultraviolet detection (HPLC-UV) method.[59] Hosotsubo et al. studied the analytic performance of mycophenolic acid EMIT and observed no interference from the major metabolite mycophenolic acid glucuronide. In addition, EMIT also correlated well with HPLC-UV method for determination of mycophenolic acid.[60] In another study the authors investigated analytic performance of Roche total mycophenolic acid assay for application on cobas Integra and cobas 6000 analyzer by comparing these methods with a specific LC-MS/MS method for determination of mycophenolic acid in specimens obtained from liver transplant recipients. The authors did not observe any significant bias between values obtained by Roche total mycophenolic acid immuno-assay and values determined by LC-MS/MS. According to the Passing-Bablok regression analysis, cobas Integra (mg/L) = 1.02 × LC-MS/MS (mg/L) − 0.50 and cobas 6000 (mg/L) = 0.98 × LC-MS/MS (mg/L) − 0.47. Due to excellent correlation with LC-MS/MS values, the authors concluded that the Roche immunoassay is suitable for therapeutic drug monitoring of mycophenolic acid.[61] However, Westley et al. observed significant bias in mycophenolic acid values determined by another immunoassay (CEDIA on Hitachi 911 analyzer) and by a chromatographic method (HPLC-UV). The regression analysis of samples from transplant recipients gave an equation: CEDIA (mg/L) = 1.18 × HPLC-UV (mg/L) + 0.45, indicating an average positive bias of 18% with the CEDIA results compared with the HPLC-UV results. The package insert indicated 192% cross-reactivity with the mycophenolic acid glucuronide metabolite. The authors concluded that mycophenolic acid CEDIA (Thermo Fisher, formerly Microgenics) run on the Hitachi 911 analyzer overestimated plasma mycophenolic acid values that are influenced by the transplant type.[62]

CONCLUSIONS

Immunoassays are widely used for routine therapeutic drug monitoring in clinical laboratories. Although most immunoassays are robust, digoxin immunoassays and immunoassays used for monitoring immunosuppressant drugs are subjected to significant interferences. Significant cross-reactivity with drug metabolite is a significant problem in therapeutic drug monitoring of cyclosporine, tacrolimus, sirolimus, and everolimus using immunoassays. However, some immunoassays for mycophenolic acid are virtually free from metabolite cross-reactivity. Although chromatographic methods, especially LC-MS/MS, are considered as the gold standard for therapeutic drug monitoring of immunosuppressants, because of their initial high cost and longer turnaroundtime, majority of hospital laboratories adopt immunoassays for therapeutic drug monitoring of immunosuppressant drugs.

REFERENCES

1. Watson I, Potter J, Yatscoff R, Fraser A, et al. *Ther Drug Monit [Editorial]*. 1997;19:125.
2. Ratanajamit C, Kaewpibal P, Setthawacharavanich S, Faroongsarng D. Effect of pharmacist participation in the health care team on therapeutic drug monitoring utilization for antiepileptic drugs. *J Med Assoc Thai*. 2009;92:1500–1507.
3. Bertholf R, Johannsen L, Bazooband A, Mansouri V. False-positive acetaminophen results in a hyperbilirubinemic patient. *Clin Chem*. 2003;49:695–698.
4. Polson J, Wians FH, Orsulak P, Fuller D, et al. False positive acetaminophen concentrations in patients with liver injury. *Clin Chim Acta*. 2008;391:24–30.
5. Fong BM, Siu TS, Tam S. Persistently increased acetaminophen concentrations in a patient with acute liver failure. *Clin Chem*. 2011;57:9–11.
6. Dasgupta A, Zaldi S, Johnson M, Chow L, et al. Use of fluorescence polarization immunoassay for salicylate to avoid positive/negative interference by bilirubin in the Trinder salicylate assay. *Ann Clin Biochem*. 2003;40:684–688.
7. Lampon N, Pampin E, Tutor JC. Investigation of possible interference by digoxin-like immunoreactive substances on the Architect iDigoxin CMIAin serum samples from pregnant women, and patients with liver disease, renal insufficiency, critical illness and kidney and liver transplant. *Clin Lab*. 2012;58:1301–1304.
8. Steimer W, Muller C, Eber B. Digoxin assays: frequent, substantial and potentially dangerous interference by spironolactone, canrenone and other steroids. *Clin Chem*. 2002;48:507–516.
9. Dasgupta A, Saffer H, Wells A, Datta P. Bidirectional (positive/negative) interference of spironolactone, canrenone, and potassium canrenoate on serum digoxin measurement: elimination of interference by measuring free digoxin or using a chemiluminescent assay for digoxin. *J Clin Lab Anal*. 2002;16:172–177.
10. DeFrance A, Armbruster D, Petty D, Cooper KC. Abbott ARCHITECT clinical chemistry and immunoassay systems: digoxin assays are free of interferences from spironolactone, potassium canrenoate, and their common metabolite canrenone. *Ther Drug Monit*. 2011;33:128–131.
11. Dasgupta A, Johnson MJ, Sengupta TK. Clinically insignificant negative interferences of spironolactone, potassium canrenoate, and their common metabolite canrenone in new dimension vista LOCI digoxin immunoassay. *J Clin Lab Anal*. 2012;26:143–147.
12. Tzou MC, Reuning RH, Sams RA. Quantitation of interference in digoxin immunoassay in renal, hepatic and diabetic disease. *Clin Pharm Ther*. 1997;61:429–441.
13. Liendo C, Ghali JK, Graves SW. A new interference in some digoxin assays: anti-murine heterophilic antibodies. *Clin Pharmacol Ther*. 1996;60:593–598.
14. Hermida-Cadahía EF, Calvo MM, Tutor JC. Interference of circulating endogenous antibodies on the Dimension® DGNA digoxin immunoassay: elimination with a heterophilic blocking reagent. *Clin Biochem*. 2010;43:1475–1477.
15. McMillin GA, Qwen W, Lambert TL, De B, et al. Comparable effects of DIGIBIND and DigiFab in thirteen digoxin immunoassays. *Clin Chem*. 2002;48:1580–1584.
16. Fyer F, Steimer W, Muller C, Zilker T. Free and total digoxin in serum during treatment of acute digoxin poisoning with Fab fragments; Case study. *Am J Crit Care*. 2010;19:387–391.

17. Chow L, Johnson M, Wells A, Dasgupta A. Effect of the traditional Chinese medicine Chan Su, Lu-Shen-Wan, DanShen and Asian ginseng on serum digoxin measurement by Tina-Quant (Roche) and Synchron LX system (Beckman) digoxin immunoassays. *J Clin Lab Anal.* 2003;17:22–27.

18. Haynes BE, Bessen HA, Wightman WD. Oleander tea: herbal draught of death. *Ann Emerg Med.* 1985;14:350–353.

19. Dasgupta A, Datta P. Rapid detection of oleander poisoning using digoxin immunoassays: comparison of five assays. *Ther Drug Monit.* 2004;26:658–663.

20. Dasgupta A, Welsh KJ, Hwang SA, Johnson M. Bidirectional (negative/positive) interference of oleandrin and oleander extract on a relatively new Loci digoxin assay using Vista 1500 analyzer. *J Clin Lab Anal.* 2014;28:16–20.

21. Alexandre J, Foucault A, Coutance G, Scanu P, et al. Digitalis intoxication induced by an acute accidental poisoning of lily of the valley. *Circulation.* 2012;125:1053–1055.

22. Edgerton PH. Symptoms of digitalis like toxicity in a family after accidental ingestion of lily of the valley plant. *J Emerg Nurs.* 1989;15:220–223.

23. Haass LF. Convallari majalis (lily of the valley) (also known as our lady's tears, ladder to heaven). *J Neurol Neurosurg Psychiatry.* 1995;59:367.

24. Welsh KJ, Huang RS, Actor JK, Dasgupta A. Rapid detection of the active cardiac glycoside convallatoxin of the lily of the valley using LOCI digoxin assay. *Am J Clin Pathol.* 2014;142:307–312.

25. Everett JM, Kojima YA, Davis B, Dasgupta A. The iDigoxin is more sensitive than LOCI digoxin assay for rapid detection of convallatoxin, the active cardiac glycoside of lily of the valley. *Ann Clin Lab Sci.* 2015;45:323–326.

26. Fink SL, Robey TE, Tarabar AF, Hodsdon ME. Rapid detection of convallatoxin using five digoxin immunoassays. *Clin Toxicol.* 2014;52:659–663.

27. Bigger Jr JT, Leahey Jr EB. Quinidine and digoxin: an important interaction. *Drugs.* 1982;24:229–239.

28. Parant F, Bossu H, Gagnieu MC, Lardet G. Cross-reactivity assessment of carbamazepine-10,11-epoxide, oxcarbazepine, and 10-hydroxy-carbazepine in two automated carbamazepine immunoassays: PETINIA and EMIT 2000. *Ther Drug Monit.* 2003;25:41–45.

29. Parant F, Moulsma M, Gagnieu MV, Lardet G. Hydroxyzine and metabolite as a source of interference in carbamazepine particle-enhanced turbidimetric inhibition immunoassay (PETINIA). *Ther Drug Monit.* 2005;27:457–462.

30. Dasgupta A, Tso G, Johnson M, Chow L. Hydroxyzine and cetirizine interfere with the PENTINA carbamazepine assay but not with the ADVIA CENTAUR carbamazepine assay. *Ther Drug Monit.* 2010;32:112–115.

31. Annesley T, Kurzyniec S, Nordblom G, et al. Glucuronidation of prodrug reactive site: isolation and characterization of oxymethylglucuronide metabolite of fosphenytoin. *Clin Chem.* 2001;46:910–918.

32. Hirata K, Saruwatari J, Enoki Y, Iwata K, et al. Possible false-negative results on therapeutic drug monitoring of phenytoin using a particle enhanced turbidimetric inhibition immunoassay in a patient with a high level of IgM. *Ther Drug Monit.* 2014;36:553–555.

33. Simons SA, Molinelli AR, Sobhani K, Rainey PM, Hoofnagle A. Two cases with unusual vancomycin measurements. *Clin Chem.* 2009;55:578–582.

34. Florin L, Vantilborgh A, Pauwels S, Vanwynsberghe T, et al. IgM interference in the Abbott iVanco immunoassay: a case report. *Clin Chim Acta.* 2015;447:32–33.

35. LeGatt DF, Blakney GB, Higgins TN, Schnabl KL, et al. The effect of paraproteins and rheumatoid factor on four commercial immunoassays for vancomycin: implications for laboratorians and other health care professionals. *Ther Drug Monit.* 2012;34:306–311.

36. Gunther M, Saxinger L, Gray M, Legatt D. Two suspected cases of immunoglobulin-mediated interference causing falsely low vancomycin concentrations with the Beckman PETINIA method. *Ann Pharmacother.* 2013;47. e19.

37. Dimeski G, Bassett K, Brown N. Paraprotein interference with turbidimetric gentamicin assay. *Biochem Med.* 2015;25:117–124.

38. Adamczyk M, Fishpaugh JR, Harrington CA, Hartter DE, et al. Immunoassay reagents for psychoactive drugs. Part 3. Removal of phenothiazine interferences in the quantification of tricyclic antidepressants. *Ther Drug Monit.* 1993;15:436–439.

39. Saidinejad M, Law T, Ewald MB, et al. Interference by carbamazepine and oxcarbazepine with serum- and urine-screening assays for tricyclic antidepressants. *Pediatrics.* 2007;120:e504–509.

40. Dasgupta A, McNeese C, Wells A. Interference of carbamazepine and carbamazepine 10,11-epoxide in the fluorescence polarization immunoassay for tricyclic antidepressants: estimation of the true tricyclic antidepressant concentration in the presence of carbamazepine using a mathematical model. *Am J Clin Pathol.* 2004;121:418–425.

41. Van Hoey NM. Effect of cyclobenzaprine on tricyclic antidepressant assays. *Ann Pharmacother.* 2005;39:1314–1317.

42. Dasgupta A, Wells A, Datta P. False-positive serum tricyclic antidepressant concentrations using fluorescence polarization immunoassay due to the presence of hydroxyzine and cetirizine. *Ther Drug Monit.* 2007;29:134–139.

43. Caravati EM, Juenke JM, Crouch BI, Anderson KT. Quetiapine cross-reactivity with plasma tricyclic antidepressant immunoassays. *Ann Pharmacother.* 2005;39:1446–1449.

44. Sorisky A, Watson DC. Positive diphenhydramine interference in the EMIT-st assay for tricyclic antidepressants in serum. *Clin Chem.* 1986;32:715.

45. Wallemacq P, Maine GT, Berg K, Rosiere T, et al. Multisite analytical evaluation of Abbott Architect cyclosporine assay. *Ther Drug Monit.* 2010;32:145–151.

46. Soldin SJ, Hardy RW, Wians FH, Balko JA, et al. Performance evaluation of the new ADVIA Centaur system cyclosporine assay (single-step extraction). *Clin Chim Acta.* 2010;411:806–811.

47. De Jonge H, Geerts I, Declercq P, de Loor H, et al. Apparent elevation of cyclosporine whole blood concentration in a renal allograft recipient. *Ther Drug Monit.* 2010;32:529–531.

48. Lee YH. Comparison between ultra-performance liquid chromatography with tandem mass spectrometry and a chemiluminescence immunoassay in the determination of cyclosporin A and tacrolimus levels in whole blood. *Exp Ther Med.* 2013;6:1535–1539.

49. Leung EK, Yi X, Gloria C, Yeo KT. Clinical evaluation of the QMS® Tacrolimus immunoassay. *Clin Chim Acta.* 2014;431:270–275.

50. Armedariz Y, Garcia S, Lopez R, Pou L, et al. Hematocrit influences immunoassay performance for the measurement of tacrolimus in whole blood. *Ther Drug Monit.* 2005;27:766–769.

51. Westley IS, Taylor PJ, Salm P, Morris RG. Cloned enzyme donor immunoassay tacrolimus assay compared with high-performance liquid chromatography-tandem mass spectrometry in liver and renal transplant recipients. *Ther Drug Monit.* 2007;29:584–591.

52. Barcelo-Martin B, Marquet P, Ferrer JM, Castanyer Puig B, et al. Rheumatoid factor interference in a tacrolimus immunoassay. *Ther Drug Monit.* 2009;31:743–745.

53. Altinier S, Varagnolo M, Zaninotto M, Boccagni P, et al. Heterophilic antibody interference in a non-endogenous molecule assay: an apparent elevation in the tacrolimus concentration. *Clin Chim Acta.* 2009;402:193–195.

54. Knorr JP, Grewal KS, Balasubramanian M, Zaki R, et al. Falsely elevated tacrolimus levels caused by immunoassay interference secondary to beta-galactosidase antibodies in an infected liver transplant recipient. *Pharmacotherapy.* 2010;30:954.

55. Rostaing L, Cointault O, Marquet P, Josse AG, et al. Falsely elevated whole blood tacrolimus concentrations in a kidney transplant patient: potential hazards. *Transpl Int.* 2010;23:227–230.

56. D'Alessandro M, Mariani P, Mennini G, Severi D, et al. Falsely elevated tacrolimus concentrations measures using the ACMIA method due to circulating endogenous antibodies in a kidney transplant recipient. *Clin Chim Acta.* 2011;412:245–248.

57. Schmidt RW, Lotz J, Schweigert R, Lackner K, et al. Multisite analytical evaluation of a chemiluminescent magnetic microparticle immunoassay (CMIA) for sirolimus on the Abbott ARCHITECT analyzer. *Clin Biochem.* 2009;42:1543–1548.

58. Dasgupta A, Davis B, Chow L. Evaluation of QMS everolimus assay using Hitachi 917 analyzer: comparison with liquid chromatography/mass spectrometry. *Ther Drug Monit.* 2011;33:149–154.

59. Dasgupta A, Tso G, Chow L. Comparison of mycophenolic acid concentrations determined by a new PETINIA assay on the dimension EXL analyzer and a HPLC-UV method. *Clin Biochem.* 2013;46:685–687.

60. Hosotsubo H, Takahara S, Imamura R, Kyakuno M, et al. Analytical validation of the enzyme multiplied immunoassay technique for the determination of mycophenolic acid in plasma from renal transplant recipients compared with a high performance liquid chromatographic assay. *Ther Drug Monit.* 2001;23:669–674.

61. Decavele AS, Favoreel N, Heyden FV, Verstraete AG. Performance of the Roche total mycophenolic acid assay on the Cobas Integra 400, Cobas 6000 and comparison to LC-MS/MS in liver transplant patients. *Clin Chem Lab Med.* 2011;49:1159–1165.

62. Westley IS, Ray JE, Morris RG. CEDIA mycophenolic acid assay compared with HPLC-UV in specimens from transplant recipients. *Ther Drug Monit.* 2006;28:632–636.

CHAPTER 7

Issues of Interference in Drugs of Abuse Testing and Toxicology

INTRODUCTION

In toxicology laboratories, therapeutic drug monitoring, serum acetaminophen and salicylate measurement in patients suspected with overdose, blood alcohol measurement, and urine drug screenings are usually conducted. In addition, heavy metals such as lead, arsenic, and mercury are tested in toxicology laboratories. However, forensic laboratories and reference laboratories have very extensive test menus.

Immunoassays are used for therapeutic drug monitoring and also screening for presence of drugs in urine, while enzymatic assay is used for serum or plasma alcohol measurements. Interferences in immunoassays used for therapeutic drug monitoring are discussed in detail in Chapter 6. In this chapter, issues of interferences in immunoassays for drugs of abuse testing and interferences in enzymatic alcohol assay as well as acetaminophen and salicylate assays are discussed. Heavy metals are tested using atomic absorption or inductively coupled plasma mass spectrometry. In addition, forensic laboratories and reference laboratories use gas chromatography/mass spectrometry (GC/MS), liquid chromatography combined with mass spectrometry (LC-MS) or LC combined with tandem MS (LC-MS/MS), and high-resolution mass spectrometry capable of screening for the presence many drugs in serum, urine, or other biological matrix. Discussion on chromatographic methods for drug analysis is beyond the scope of this chapter.

Drug testing may be medical or legal but most drug tests conducted in clinical laboratories are for medical purpose only. For medical drug testing, urine specimens are analyzed using immunoassays for the presence of any prescription or illicit drugs. However, positive screening results may not be confirmed using a chromatographic method such as gas GC/MS, high-performance liquid chromatography (HPLC),LC-MS, or LC-MS/MS. In contrast, for legal drug testing, for example, workplace drug testing, drug confirmation by an alternative method, preferably GC/MS or LC-MS/MS, is mandatory. Moreover, chain of custody must be maintained for the specimen analyzed and a medical review officer (MRO) must

review the result to ensure that there is no alternative explanation for the positive test result. For example, a person taking prescription oxycodone will show positive oxycodone in urine confirmed by a chromatographic method. This is known as analytic positive. However, after reviewing results and contacting the clinician to confirm oxycodone prescription, the MRO may determine that this person should be hired because the individual is taking oxycodone under medical supervision.

DRUGS AS CONTROLLED SUBSTANCES

In most countries, drugs with high abuse potential are strictly regulated by the government. In the United States, The Drug Abuse Control Act of 1956 provided guidelines for pharmaceutical industries for manufacturing and dispensing controlled substances. Then in 1970, the Controlled Substances Act was passed for further regulating drugs with high abuse potential. The major focus of this law was the scheduling of drugs into five different classes based on abuse potential, harmfulness, and development of drug dependence, as well as potential benefits when used medically. Several amendments were later added in order to provide power to the Attorney General of the United States and subsequently to the Drug Enforcement Administration (DEA) to classify a drug with high abuse potential in "Schedule I" prior to completion of formal review. The two other well-known amendments to the drug act were Anti-Drug Abuse Acts of 1986 and 1988.[1] Controlled substances are categorized in five groups depending on the medical need and abuse potential. Schedule I drugs have no known medical use but very high abuse potential. Schedule II drugs may have known medical use but are also highly addictive. Schedule III drugs are used medically but may have some abuse potential. However, abuse potential of Schedule III drugs are less in magnitude than Schedule II drugs. Schedule IV drugs are used in medical practice but may have low potential of abuse. Schedule V drugs are widely used in medical practice and some drugs may have very low abuse potential. Example of a Schedule V drug is cough mixture containing low level of codeine.

Biotin and Other Interferences in Immunoassays. https://doi.org/10.1016/B978-0-12-816429-7.00007-1

WORKPLACE DRUG TESTING

The workplace drug testing was initiated by President Reagan who issued the executive order number 12564 on September 15, 1986. This executive order directed drug testing for all federal employees who are involved in law enforcement, national security, protection of life and property, public health, and other services requiring high degree of public trust. Following this executive order the National Institute on Drug Abuse was given the responsibility of developing guidelines for federal drug testing. Currently, Substance Abuse and Mental Health Services Administration (SAMHSA), affiliated with the Department of Health and Human Services of the Federal Government, is responsible for providing mandatory guidelines for federal workplace drug. Bush[2] summarized the guidelines for legal drug testing.

WHICH DRUGS ARE TESTED?

Traditionally in Federal drug testing, five SAMHSA mandated drugs including amphetamine, cocaine (tested as benzoylecgonine), opiates, phencyclidine (PCP), and marijuana tested as 11-nor-9-carboxy-Δ^9-tetrahydrocannabinol have been tested. In the 2015 revision to the proposed guidelines, SAMHSA recommended additional testing for oxycodone, oxymorphone, hydrocodone, and hydromorphone.[3] Some private employers may test for additional drugs in their workplace drug testing protocols and such comprehensive drug panel may include barbiturates, benzodiazepines, methadone, methaqualone, propoxyphene, fentanyl, and lysergic acid diethylamide (LSD).

For drug testing in urine, either the parent drug or its metabolite is targeted. For SAMHSA-mandated drugs, recommended cutoff concentrations for both immunoassay and GC/MS of various drugs are available. In general, such guidelines are also followed in medical drug testing. Usually a drug or its metabolites can only be detected in urine for a limited time after last abuse. However, detection time may vary also on the dose administered as well as characteristics of screening and confirmation assay. Screening and confirmation cutoff and the window of detection of common drugs that are tested in both medical and legal drug testing are listed in Table 7.1.

Analytic Methods

In general the first step in drug testing (both medical and legal) is screening for the presence of any drug using FDA-approved immunoassays. Unless specifically requested by a clinician for drug confirmation, in medical drug testing, screening-positive specimens are not confirmed by GC/MS or LC-MS/MS. However, in legal drug testing, confirmation test is mandatory because false-positive immunoassay screening results are common, especially with amphetamine/methamphetamine screening assays. Moreover, in legal drug testing, if one specimen is screened positive for the presence of a drug/drug class but that drug cannot be confirmed by GC/MS or LC-MS/MS, the result must be reported as "negative."

If a drug testing result is negative, it does not exclude abuse of any illicit drugs. Every drug has a window of detection in urine drug testing. Moreover, many designer drugs including bath salts (synthetic cathinone) and spices (K2 Blonde, etc.; synthetic cannabinoids) cannot be detected during routine drug screen because bath salts, despite having some structural similarity with amphetamine, may not have enough cross-reactivity to test positive and synthetic cannabinoids cannot be detected by marijuana immunoassays (targeting inactive metabolite 11-nor-Δ^9-tetrahydrocannabinol-9-carboxylic acid, THC-COOH), as they are not structurally similar to marijuana (tetrahydrocannabinol). These designer drugs are called "synthetic cannabinoids" because their pharmacologic actions are due to interaction with cannabinoid receptors (CB1 and CB2) in the brain.

Major interferences in drug testing are observed during immunoassay screening of drugs. This chapter addresses these issues. GC/MS and LC-MS/MS are relatively free from interference. However, there are issues of ion suppression, isobaric ions, and proper selection of the internal standard (most desirable internal standard is deuterated analog of the analyte) in chromatographic methods, which may impact test results. If proper attention is focused on these analytic issues during development of chromatography-based methods, then such assays should be virtually free from interference.

INTERFERENCES IN AMPHETAMINE IMMUNOASSAYS

Immunoassays for amphetamines are also capable of detecting 3,4-methylenedioxyamphetamine (MDA) and 3,4-methylenedioxymethamphetamine (MDMA). However, certain amphetamine immunoassays may have lower capability of detecting MDMA and MDA due to lower cross-reactivity of the antibody used for MDMA and MDA. Poklis et al. reported that the EMIT d.a.u monoclonal amphetamine/methamphetamine immunoassay has a cutoff concentration of 3000 ng/mL for racemic MDMA but only 800 ng/mL for MDA.

TABLE 7.1
Screening, Confirmation Cutoff, and Window of Detection of Various Drugs in Urine

	Target Analyte in Urine	Window of Detection	Screening Cutoff (ng/mL)	Confirmation Cutoff (ng/mL)
SAMHSA DRUGS				
Amphetamine/ methamphetamine	Amphetamine/ methamphetamine	2 days	500	250
MDMA/MDA/MDEA	MDMA	2 days	500	250 ng/mL for MDMA/ MDA/MDEA
Cocaine	Benzoylecgonine	2–4 days	150	100
Opiates (morphine/ codeine)	Morphine	3 days	2000	2000 ng/mL for either drug
Heroin	6-Monoacetyl-morphine	12 h to 1 day	10	10
Hydrocodone/ hydromorphone	Hydrocodone	3 days	300	100 ng/mL for either drug
Oxycodone/ oxymorphone	Oxycodone	3 days	100	50 ng/mL for either drug
Marijuana	THC-COOH	2–3 days/Single 30 days/Chronic	50	15
PCP	PCP	8 days	25	25
NON-SAMHSA DRUGS				
Barbiturates	Secobarbital	3 days/Short acting 15 days/Long acting	200/300	200
Benzodiazepines	Oxazepam/nordiazepam and others	2 days/Short acting 10 days/Long acting	200/300	200
Propoxyphene	Propoxyphene	3 days	300	300
Methadone	Methadone or EDDP	3 days	300	100 ng/mL

EDDP, 2-ethylidene-1,5-dimethyl-3,3-diphenylpyrrolidine; *MDA*, 3,4-methylenedioxyamphetamine; *MDEA*, 3,4-methylenedioxyethylamphetamine; *MDMA*, 3,4-methylenedioxymethamphetamine; *SAMHSA*, Substance Abuse and Mental Health Services Administration; *THC-COOH*, 11-nor-Δ^9-tetrahydrocannabinol-9-carboxylic acid.

The assay had higher sensitivity for detecting the S(+) isomer of both MDMA and MDA.[4] The Roche Abuscreen ONLINE amphetamine immunoassay also has low cross-reactivity toward MDMA but higher cross-reactivity with MDA.[5] However, MDMA is metabolized to MDA and this assay may be able to identify individuals abusing MDMA. In addition, specific immunoassays are commercially available where the assay antibody targets MDMA molecules.

Stout et al. studied the performances of four immunoassays (DRI amphetamine, DRI ecstasy, Abuscreen ONLINE amphetamines, and a modified Abuscreen ONLINE amphetamine) for detection of amphetamine, methamphetamine, MDA, and MDMA. The modified ONLINE reagent was calibrated with MDMA and had 16 mM sodium periodate added to the R2 reagent.

These assays were run on approximately 27,500 human urine samples and 7000 control urine samples prepared at 350 and 674 ng/mL, respectively, over the course of 8 days. GC-MS confirmation was conducted on screened-positive samples. The authors reported that the DRI ecstasy reagent provided improved sensitivity for MDMA as compared with the ONLINE reagent, with approximately 23% more samples screening and confirming positive for MDMA and a confirmation rate of approximately 90%.[1]

Amphetamine immunoassays may not be sensitive to detect amphetamine-like designer drugs, including bath salts. Kerrigan et al. evaluated cross-reactivities of 11 designer drugs, with 9 various commercially available immunoassays. The 11 designer drugs included in the study were 2,5-dimethoxy-4-bromophenethylamine

(2C-B); 2,5-dimethoxyphenethylamine (2C-H); 2,5-dimethoxy-4-iodophenethylamine (2C-I); 2,5-dimethoxy-4-ethylthiophenethylamine (2C-T-2); 2,5-dimethoxy-4-isopropylthiophenethylamine (2C-T-4); 2,5-dimethoxy-4-propylthiophenethylamine (2C-T-7); 2,5-dimethoxy-4-bromoamphetamine (DOB); 2,5-dimethoxy-4-ethylamphetamine (DOET); 2,5-dimethoxy-4-iodoamphetamine(DOI); 2,5-dimethoxy-4-methylamphetamine (DOM); and 4-methylthioamphetamine (4-MTA). Cross-reactivities of these designer drugs with immunoassays studied were <0.4%, and even at a concentration of 50,000 ng/mL, these designer drugs were insufficient to produce a positive response. However, 4-MTA was the only drug that demonstrated 5% cross-reactivity with Neogen amphetamine enzyme-linked immunosorbent assay (ELISA) (Lexington, KY) but a significant 200% cross-reactivity with Immunalysis amphetamine ELISA (Pomona, CA).[6]

Apollonio et al.[7] reported that 4-MTA had 280% cross-reactivity with the Bio-Quant Direct ELISA for amphetamine. Petrie et al. concluded that some amphetamine-like designer drugs at concentrations of 20,000 ng/mL may produce positive result with various amphetamine immunoassays but existing immunoassays unevenly detect amphetamine-like designer drugs, particularly the 2C series, piperazine, and β-keto class drugs (synthetic cathinone; bath salts).[8]

Currently, there is a commercially available immunoassay (Randox) that can detect certain bath salts if present in the urine specimen. The Randox Drugs of Abuse V (DOA-V) Biochip Array Technology contains two synthetic cathinone antibodies: Bath Salt I antibody targets mephedrone/methcathinone and Bath Salt II antibody targets 3′,4′-methylenedioxypyrovalerone/3′,4′-methylenedioxy-α-pyrrolidinobutiophenone. In one study the authors observed that if the pH of urine was less than 4, false-positive test results may be obtained. The authors also analyzed 20,017 urine specimens but observed limited confirmed-positive specimens.[9]

Amphetamine and methamphetamine have optical isomers designated *d* (or +) for dextrorotatory and *l* (or −) for levorotatory. Amphetamine immunoassay antibody targets *d*-isomer because only the *d*-isomer is abused for its pharmacologic effect. The Vicks inhaler contains the *l*-methamphetamine. Legitimate use of legal intranasal decongestants containing *l*-methamphetamine may complicate interpretation of amphetamine immunoassay test results. Poklis et al. reported relatively high concentrations of *l*-methamphetamine in two subjects (1390 ng/mL and 740 ng/mL) after extensively inhaling Vicks inhaler every hour for several hours. However, urine specimens tested negative

by the enzyme multiplied immunoassay technique (EMIT) II amphetamine assay even after such extensive use of Vicks inhaler.[10] In contrast, in another study the authors observed 2.2% false-positive test results using EMIT II Plus amphetamine assay when volunteers extensively used Vicks VapoInhaler. In the study design, 22 healthy adults were each administered one dose (two inhalations in each nostril) from a Vicks VapoInhaler every 2 h for 10 h on day 1 (six doses), followed by a single dose on day 2. Every urine specimen was collected as an individual void for 32 h after the first dose and assayed for *d*- and *l*-amphetamines using GC/MS after chiral derivatization to form R(−)-α-methoxy-α-(trifluoromethyl)phenylacetyl derivatives. The median *l*-methamphetamine maximum concentration was 62.8 ng/mL (range, 11.0–1440) and only two subjects had detectable *l*-amphetamine levels. The authors concluded that after extensive use of Vicks inhaler, it is possible to test positive using the amphetamine immunoassay, requiring an enantiomer-specific confirmation.[11]

The major problem of amphetamine immunoassay is false-positive test results after using many over-the-counter cold medications containing ephedrine or pseudoephedrine. Although cross-reactivities are relatively low (0.4%–2%) with assay antibody, false-positive test results are common after taking over-the-counter cold medication containing ephedrine or pseudoephedrine because urinary concentrations of these compounds are high enough to cause false-positive test results at 500 ng/mL cutoff concentration of amphetamine/methamphetamine. In one study the authors reported that after ingesting a therapeutic dose of pseudoephedrine (60 mg), urinary pseudoephedrine concentration may exceed 100,000 ng/mL in some subjects, while the concentration of cathine (metabolite of pseudoephedrine) may exceed 5000 ng/mL. The authors also observed a high interindividual variability in the urinary concentration of ephedrine/pseudoephedrine after the administration of the same therapeutic dose of a preparation.[12] DePriest et al.[13] commented that a false-positive rate of 75% was reported with the Microgenics (now Thermo Fisher) DRI amphetamine assay due to the presence of pseudoephedrine.

Ranitidine is a H$_2$ receptor–blocking agent (antihistamine) that reduces acid production by the stomach and is available over the counter without any prescription. Dietzen et al. reported that ranitidine, if present in urine at a concentration over 43 μg/mL, may cause a false-positive amphetamine screen test result using Beckman SYNCHRON immunoassay reagents (Beckman Diagnostics, Brea, CA). This concentration of ranitidine is expected in patients taking ranitidine at the recommended therapeutic dose.[14] However, ranitidine does not interfere with the Siemens EMIT II Plus assay. In one study the authors reported that ranitidine at 160 μg/mL can produce false-positive results on the Beckman Coulter amphetamine assay with a 1000 ng/mL cutoff, although the manufacturer states that a positive amphetamine result would not be produced in samples with 1000 μg/mL ranitidine. On the other hand, ranitidine concentrations up to 5000 μg/mL failed to produce positive results on the Siemens EMIT II Plus assay with a 1000 ng/mL amphetamine calibrator cutoff.[15]

Casey et al. reported that bupropion, a monocyclic antidepressant and an aid for smoking cessation, may cause false-positive screen results using the EMIT II amphetamine immunoassay. The authors conducted an IRB (Institutional Review Board)-approved retrospective chart review of all emergency department patients who underwent urine drug screen between January 1, 2006, and July 31, 2007, using Syva EMIT II Plus immunoassay reagents. All positive screens also underwent confirmation. Of the 10,011 urine drug screens, 362 (3.6%) were positive for amphetamine by the immunoassay, but the presence of amphetamines was not confirmed in 128 (35%) out of these 362 amphetamine screen-positive urine specimens. Among the 128 urine specimens where the presence of amphetamines was not confirmed, records reflected prescription use of bupropion in 53 (41%). The authors concluded that the therapeutic use of bupropion may produce false-positive amphetamine test results.[16]

Logan et al. reported that a series of patients whose urine screened positive for MDMA using a commercially available Ecstasy EMIT II assay tested negative for MDMA using a specific LC-MS/MS method for confirming MDMA. Further evaluation of these urine specimens indicated that these specimens were positive for trazodone and its metabolite meta-chlorophenylpiperazine (m-CPP). Testing with standard compounds showed significant cross-reactivity of trazodone, m-CPP, and the related recreational drug trifluoromethyl phenylpiperazine with the Ecstasy EMIT II assay. The authors commented that their findings have further forensic significance because m-CPP is emerging as an illicit recreational drug in its own right or as an adulterant in illicit cocaine and MDMA. The hallucinogen benzylpiperazine was also assessed but found not to cross-react significantly with this assay.[17]

Baron et al. also reported false-positive amphetamine test results using Amphetamines II immunoassay (Roche Diagnostics) in patients taking trazodone.

The authors tested real patient urine samples containing m-CPP (detected and quantified by HPLC) with no detectable amphetamine, methamphetamine, or MDMA (demonstrated by GC/MS) and found that in patients taking trazodone the urine may have sufficient m-CPP to cause false-positive Amphetamines II results. The authors also showed significant cross-reactivity of m-CPP with this amphetamine immunoassay using the standard material. The authors further commented that at their institution, false-positive amphetamine results occur not infrequently in patients taking trazodone, with at least eight trazodone-associated false-positive results during a single 26-day period. Laboratories should remain cognizant of this interference when interpreting results of this assay.[18]

Labetalol, a β-blocker commonly used for control of hypertension in pregnancy, can cause false-positive amphetamine screen results using an immunoassay. An oxidative metabolite of labetalol, 3-amino-1-phenylbutane, was initially identified in a patient's urine by GC/MS responsible for cross-reactivity with amphetamine assays. This metabolite was shown to cross-react approximately 2% with the Abbott TDx amphetamine/methamphetamine II kit, 10% with the Syva EMIT d.a.u. polyclonal amphetamine class kit, and 3% with the Syva EMIT d.a.u. monoclonal amphetamine kit. This degree of cross-reactivity is sufficient to cause false-positive amphetamine test results.[19]

Case Report

A 38-year-old gravida 2, para 0 woman was transferred to the hospital at 28 weeks of gestation. At a routine appointment 2 days before admission, she was found to have severe hypertension, pitting edema, hyperreflexia, and proteinuria. She also showed persistent systolic blood pressures more than 170 mmHg. She was diagnosed of superimposed preeclampsia on chronic hypertension. She received betamethasone for fetal lung maturity, magnesium for seizure prophylaxis, and labetalol for blood pressure control (800 mg three times daily). A urine drug test was positive for amphetamine, but as the patient was hospitalized for more than 48 h before the urine drug screen, there was no evidence of access to amphetamines under direct inpatient observation. Thus confirmatory testing was not performed, given the likely false-positive result because of labetalol administration in this patient.[20]

There is one case report of a false-positive amphetamine immunoassay screening result in a 60-year-old diabetic patient who was on the antidiabetic medication metformin for the past 19 years. No amphetamine was detected during confirmation.[21] The interference of buflomedil, a peripheral vasodilator with the monoclonal EMIT d.a.u. amphetamine immunoassays, has been reported. Urine samples collected from 20 patients taking 600 mg of buflomedil daily showed false-positive results with the monoclonal EMIT d.a.u. assay, as did urine specimens collected 2 h after the first oral dose of buflomedil. The authors commented that one or more buflomedil metabolites, besides the unchanged drug, probably interfere in the monoclonal EMIT d.a.u. assay.[22] In one study the authors evaluated potential interference of mexiletine, a class IB antiarrhythmic drug, with several amphetamine screening assays: the Syva EMIT II Plus and the Roche Kinetic Interaction of Microparticles in Solution (KIMS) automated immunoassays, along with the Noble Split-Specimen and SYNCHRONQuikScreen point-of-care assays. Urine samples from two patients treated with mexiletine were positive on all amphetamine screens but confirmed negative by GC/MS. Drug-free urine spiked with mexiletine caused positive results on all assays, although the EMIT II Plus and KIMS assays cross-reacted at lower mexiletine concentrations than the point-of-care assays.[23] Chloroquine, at high concentrations, may also cause positive amphetamine screening test results.

Case Report

A 14-year-old girl, after a family dispute, ingested four tablets of each containing 250-mg chloroquine diphosphate salt. She presented with mild headache and dizziness but with no other symptoms or palpitations. Her blood pressure was 116/69 mm Hg and pulse was 90 beats per minute and regular. Two urine specimens, one collected 5 h after ingestion of chloroquine and another collected 9 h after ingestion, showed positive amphetamine test results using the DRI amphetamine assay (at 1000 ng/mL and 500 ng/mL cutoff). However, confirmation test result was negative. The chloroquine urine levels were 103,900 (5 h after ingestion) and 100,900 ng/mL (9 h after ingestion). Results of in vitro study showed that chloroquine has approximately 0.89% cross-reactivity at a cutoff of 500 ng/mL in the DRI amphetamine assay. However, hydroxychloroquine showed no cross-reactivity. Samples from patients treated with chloroquine or hydroxychloroquine did not show false-positive results using the DRI amphetamine assay. The authors concluded that chloroquine can only cause positive assay result if present in high amount in urine.[24]

In another study the authors showed false-positive amphetamine test results using amphetamine cloned enzyme donor immunoassay (CEDIA) due to high concentration of chloroquine in urine.[25] Amphetamines have an extensive list of cross-reacting medications including antidepressants and antipsychotics. In one report the authors presented two cases where urine drug screen was positive for amphetamine after ingestion of aripiprazole, an antipsychotic medication. Case 1 was a 16-month-old girl who accidently ingested 15–45 mg of aripiprazole. She was lethargic and ataxic at home with one episode of vomiting containing no identifiable tablets. Her urine specimens tested positive for amphetamines 2 days following ingestion but GC/MS confirmation was negative. Case 2 was of a 20-month-old girl who was brought into the hospital after accidental ingestion of an unknown quantity of her father's medications, which included aripiprazole. Urine drug test result on the first day of admission was positive by immunoassay but negative by GC/MS confirmation test. Eventually both patients screened negative for amphetamine and were discharged from the hospital. The authors concluded that aripiprazole may cause false-positive test results in immunoassays screening for amphetamines.[26]

The antidepressant desipramine and the antiviral agent amantadine also interfere with amphetamine immunoassays.[27] Another antidepressant doxepin is also known to cause false-positive results in amphetamine immunoassays.[28] False-positive results for urine amphetamine and opiate immunoassays in a patient intoxicated with perazine, an antipsychotic drug belonging to phenothiazine class, have also been reported.[29] Isometheptene also cross-reacts in the EMIT amphetamine assay.[30] There is one case report of false-positive amphetamine screening result due to use of methylphenidate.[31] However, Breindahl and Hindersson observed no cross-reactivity of methylphenidate causing false-positive amphetamine results in hundreds of authentic routine urine samples from patients treated with methylphenidate that were screened for amphetamines by EMIT(Siemens). The package insert also reported no cross-reactivity for methylphenidate (at 1000 ng/mL).[32]

Vorce et al. showed that 1,3-dimethylamylamine (DMAA), an aliphatic amine naturally found in geranium flowers and also used in bodybuilding natural supplements such as Jack3d and OxyElite Pro, may cause false-positive test results in the KIMS amphetamine assay (Roche Diagnostics) and the EMIT II Plus amphetamine assay if present at a concentration of 6900 ng/mL and because of its structural similarity with amphetamine. DMAA has been promoted as a safe alternative to ephedrine. The authors further analyzed 134 urine specimens that tested false positive for amphetamine but confirmed negative by GC/MS and did not contain any known drugs that may cause false-positive amphetamine test results

TABLE 7.2
Common Drugs That may Cause False-Positive Test Results With Amphetamine Immunoassay

Drug Class	Examples of Specific Drugs	Comments
Decongestants	Ephedrine, pseudoephedrine, phenylephrine	Decongestants such as ephedrine, pseudoephedrine, and phenylephrine are most commonly encountered drugs causing false-positive amphetamine result using immunoassays.
Antiarrhythmic	Mexiletine	Interferes with several amphetamine immunoassays.
Antihistamine	Brompheniramine,ranitidine	Therapeutic use of ranitidine may cause false-positive test results with BeckmanCoulter amphetamine assay but not with EMIT assay marketed by Siemens.
Antidiabetic	Metformin	There is a case report of false-positive amphetamine test result in a diabetic patient taking metformin for 10 years.
Appetite suppressant	Phentermine	False-positive results
Antidepressant	Trazodone, doxepin, desipramine, trimipramine	Trazodone cross-reactivity is due to its metabolite *m*-chlorophenylpiperazine and therapeutic use of this drug may cause false-positive test results.
Antidepressant/smoking cessation	Bupropion	Bupropion after therapeutic use may cause false-positive test results.
Antipsychotics	Aripiprazole,chlorpromazine,promethazine, perazine	False-positive test results.
β-Blocker	Labetalol	Case reports of false-positive test results due to the use of labetalol. This is due to cross-reactivity of labetalol metabolite 3-amino-1-phenylbutane with amphetamine immunoassays.
Antimalarial	Chloroquine	Positive test result if present in high concentration.
Antiviral/antiparkinsonian	Amantadine	Amantadine may cause false-positive results with amphetamine immunoassays.
Antimigraine	Isometheptene	Cross-reacts with EMIT assay.
Peripheral vasodilator	Buflomedil	Positive result with EMIT d.a.u monoclonal assay.
Herbal supplement (OxyElite Pro)	DMAA	DMAA present in herbal weight loss products may interfere with amphetamine immunoassay.

DMAA, 1,3-diemthylamylamine.

and observed the presence of DMAA in 92.3% specimens, with concentrations varying from 2500 to 67,000 ng/mL.[33]

Case Report

A 25-year-old woman presented for eligibility screening as a healthy volunteer for a mental health study. Her physical examination was unremarkable but her urine drug screen was positive for amphetamine. She stated that she was taking OxyElitePro she purchased from at a local nutrition store for weight loss and was able to lose 13.6 kg (30 lbs) over 6 months. Because she denied any illicit drug use, a confirmatory test for amphetamine was ordered, which was negative. The false-positive amphetamine screening result was due to use of herbal weight loss product OxyElitePro that contains DMAA.

The authors commented that DMAA was introduced in 1948 as a nasal decongestant but was later withdrawn from the market. However, DMAA reappeared in 2006 and is present in more than 200 supplements.[34]

Brahm commented that many drugs including brompheniramine, bupropion, chlorpromazine, clomipramine, dextromethorphan, diphenhydramine, doxylamine, ibuprofen, naproxen, promethazine, quetiapine, quinolones (ofloxacin and gatifloxacin), ranitidine, sertraline, thioridazine, trazodone, venlafaxine, verapamil, and a nonprescription nasal inhaler may cause false-positive test results in immunoassays for various drugs but false-positive results for amphetamine and methamphetamine were the most commonly reported.[35] Common drugs that may cause false-positive results in amphetamine immunoassays are listed in Table 7.2.

Case Report

A 60-year-old male patient with alcohol dependency was admitted to the inpatient psychiatry unit on his request for an alcohol treatment program. He was also diagnosed with hyperlipidemia 3 years ago and insomnia 5 months ago. He was treated with 267 mg/day fenofibrate and 100 mg/day trazodone. His complete blood count, liver and kidney function tests, thyroid profile, and urine analysis were all normal. After his admission to the psychiatry unit, fenofibrate 267 mg was administered at 6.00 p.m. and trazodone 100 mg was administered at 10.00 p.m. In addition, he was put on diazepam 7.5 mg four times daily (discontinued on day 5) to avoid alcohol withdrawal symptoms and thiamine 100 mg three times daily to supplement diet. On the second day of his admission, he tested positive for MDMA using both the DRI ecstasy assay and the amphetamine/MDMA (CEDIA) test. The second urine specimen collected on day 7 was also positive by both assays. As he had no access to outside drugs, assay interference was suspected and GC/MS confirmation was negative. As trazodone cannot be discontinued, the lipid-lowering drug fenofibrate was discontinued and the sixth sample after a 4-day fenofibrate-free interval resulted negative both for amphetamine/MDMA (CEDIA) and MDMA (DRI) tests. The authors concluded that false-positive test result was due to fenofibrate.[36]

Prescription Drugs Metabolized to Amphetamine/Methamphetamine

Amphetamine and methamphetamine are listed by the DEA as Schedule II controlled substances. Therapeutic uses of amphetamine include treatment of attention deficit disorder with hyperactivity, narcolepsy, and obesity. Methamphetamine is much less commonly used as a prescription drug, being mainly used for treatment of obesity. There are also medications that metabolize to methamphetamine or amphetamine. If a patient is on one of these medications, the positive amphetamine/methamphetamine result is a clinical false positive. For example, selegiline (Eldepryl, Zelapar), a drug used in the treatment of Parkinson disease, is metabolized to l-methamphetamine (and also l-amphetamine). Clobenzorex is a prescription drug that is metabolized to d-amphetamine.[37] Drugs containing amphetamine/methamphetamine and drugs metabolized to amphetamine or methamphetamine are listed in Table 7.3.

INTERFERENCE IN BENZOYLECGONINE IMMUNOASSAYS

Cocaine is detected in urine by identification of benzoylecgonine, an inactive metabolite of cocaine. In contrast to amphetamine immunoassays, which are subjected to interference from approximately 30 drugs, immunoassays for benzoylecgonine are robust. There is an isolated report of antibody-mediated interference in EMIT for benzoylecgonine.[38] However, true positive test results may occur due to drinking herbal tea containing cocaine or application of cocaine as an anesthetic agent during ENT surgery. In one study the authors investigated if the results of routine urine drug screening for cocaine metabolites would be positive after dacryocystorhinostomy. Postoperative urine specimens were analyzed for the presence of benzoylecgonine, the major metabolite of cocaine, using the Syva EMIT assay and confirmation using GC/MS. The results of urine tests of all 12 patients were positive for cocaine 24 h after surgery, 9 (75%) were positive 48 h after surgery, and 3 (33%) had detectable levels 72 h after surgery. The authors concluded that patients given cocaine at the time of lacrimal surgery should be warned that they may test positive for cocaine for at least 3 days after surgery.[39]

Herbal teas made from coca leaves contain cocaine. Such teas are readily available in some South American countries. Therefore drinking such tea may cause positive test result for cocaine. In one study, authors found an average 5.11 mg of cocaine per tea bag of coca tea originated from Peru and an average 4.86 mg of cocaine in per tea bag in Bolivian coca tea. When tea was prepared, one cup of Peruvian coca tea had an average of 4.14 mg of cocaine, whereas one cup of Bolivian tea had an average of 4.29 mg of cocaine. When one volunteer drank one cup of Peruvian tea, a peak benzoylecgonine concentration of 3940 ng/mL was observed 10 h after consumption. Similarly, consumption of one cup of Bolivian tea by a volunteer resulted in a peak benzoylecgonine concentration of 4979 ng/mL 3.5 h after consumption of tea. Therefore drinking such tea will cause positive results in drug testing.[40] Although the US custom regulations require that no cocaine should be present in any herbal tea, literature references indicate that some health Inca tea sold in the United States contains cocaine. Jackson et al. reported urinary concentration of benzoylecgonine after ingestion of one cup of health Inca tea by volunteers. Benzoylecgonine was detected up to 26 h after ingestion. Maximum urinary benzoylecgonine concentration ranged from 1400 to 2800 ng/mL after ingestion of health Inca tea. The total excretion of benzoylecgonine in 36 h ranged from 1.05 to 1.45 mg, which correlated with 59%–90% of the ingested cocaine dose.[41]

Coca tea or mate de coca is a commercially available tea made from coca leaves (*Erythroxylum coca*). This tea is available in South America and may also be found

TABLE 7.3	
Drugs Containing Amphetamine/Methamphetamine or Metabolized to Amphetamine/Methamphetamine	
Drug	**Comments**
Adderall	Adderall is a combination of amphetamine and dextroamphetamine and is used for treating attention deficit disorders and narcolepsy.
Amphetaminil	This drug is metabolized into amphetamine and is used for treating narcolepsy and obesity.
Benzphetamine (Didrex)	Benzphetamine, an appetite suppressant, is metabolized to methamphetamine and then into amphetamine.
Clobenzorex	Clobenzorex is an appetite suppressant that is metabolized into amphetamine.
Dimethylamphetamine	This compound is metabolized into methamphetamine. It is often found as an impurity in illicit methamphetamine preparation.
Dexedrine	Dexedrine contains dextroamphetamine and is used in treating attention deficit disorders and narcolepsy.
Dextrostat	Dextrostat contains dextroamphetamine and is used in treating attention deficit disorders and narcolepsy.
Desoxyn	Desoxyn contains d-methamphetamine and is used for treating attention deficit disorders and also obesity.
Ethylamphetamine	An older drug used as appetite suppressant is metabolized into amphetamine.
Fenethylline	Fenethylline is a theophylline derivative of amphetamine having stimulant effects similar to those of other amphetamine-type derivatives. It is metabolized into amphetamine.
Lisdexamfetamine	Lisdexamfetamine (L-lysine-dextroamphetamine) is a prodrug that is converted into d-amphetamine. It is used in treating attention deficit disorders.
Mefenorex	Mefenorex (Rondimen, Pondinil, Anexate), a stimulant drug, is used as an appetite suppressant. It is metabolized into amphetamine.
Selegiline	Selegiline is used to treat symptoms of Parkinson disease. This drug is metabolized into l-methamphetamine and then into l-amphetamine.
Vicks inhaler	A nonprescription medication containing l-methamphetamine.

in the United States. Mazor et al. studied the effect of drinking coca tea on the excretion of cocaine metabolite in urine. Five healthy volunteers consumed coca tea and underwent serial urine testing for cocaine metabolites using the fluorescence polarization immunoassay (FPIA). Each participant showed positive urine sample (over 300 ng/mL cutoff for cocaine metabolite) 2 h after drinking coca tea, and urine specimens from three out of five volunteers showed positive test results up to 36 h. Mean benzoylecgonine concentration was 1777 ng/mL (range, 1065–2495). Authors concluded that coca tea ingestion resulted in positive urine test for the cocaine metabolite.[42] Turner et al. reported positive test results for cocaine metabolite in subjects after drinking mate de coca tea. Tea was prepared by allowing one mate de coca tea bag to be immersed in 250 mL of boiling water for 25 min. The bag was removed and squeezed in tea to drain additional water. A 5-mL sample was taken for analysis and volunteers drank the rest. Urinary samples were collected at 2, 5, 8, 15, 21, 24, 43, and 68 h after drinking tea. All urine samples tested positive for benzoylecgonine, the metabolite of cocaine by immunoassay. The amount of cocaine in tea was estimated to be 2.5 mg.[43] Cocaine content in herbal teas made from coca leaves is summarized in Table 7.4.

It is important to note that life-threatening acute cocaine overdose may not be identified by urine toxicologic screen because sufficient concentration of benzoylecgonine may not be present in urine just after a severe overdose. In one case report, where the person died from cocaine overdose, the urine drug test result was negative for cocaine using EMIT. Later, GC/MS analysis confirmed that the concentration of benzoylecgonine was only 75 ng/mL in urine, which was significantly below the 300 ng/mL cutoff. However, the heart blood concentration of cocaine in the deceased was 18,330 ng/mL consistent with fatality due to cocaine overdose.[44]

TABLE 7.4
Cocaine Content in Herbal Teas Made From Coca Leaves

Herbal Tea	Cocaine Content/ Tea Bag	Cocaine Content in a Cup of Tea	Reference
Bolivian coca tea	4.86 mg	4.29 mg	40
Peruvian tea	5.11 mg	4.14 mg	40
Mate de coca	Not measured	2.5 mg	43

INTERFERENCE IN BENZODIAZEPINE IMMUNOASSAYS

Oxaprozin (Daypro) is a nonsteroidal antiinflammatory drug that cross-reacts with several benzodiazepine immunoassays. In one study the authors analyzed urine specimens obtained from 12 subjects after receiving a single oral dose (1200 mg) of oxaprozin and using 200-ng/mL cutoff for benzodiazepines; all 36 urine specimens collected from the 12 subjects gave positive results by EMIT and CEDIA, and 35 of 36 urine specimens were positive by FPIA benzodiazepine immunoassays. The authors concluded that presumptive positive benzodiazepine results by these immunoassays may be due to the presence of oxaprozin or oxaprozin metabolites and recommended that all positive immunoassay screening tests for benzodiazepines should be confirmed by another method.[45] There is a case report of positive interference of the analgesic nefopam in the urine immunoassay for benzodiazepines in a patient. After discontinuation of nefopam, benzodiazepine immunoassay result was negative.[46] Cross-reactivity of nefopam and its metabolites with benzodiazepine EMIT has also been reported.[47] Although sertraline, a commonly prescribed antidepressant, is known to interfere with benzodiazepine immunoassays, an improved CEDIA benzodiazepine assay is free from such interference.[48] Nevertheless, a study indicates that some benzodiazepine assays are still affected by sertraline. Nasky et al. performed a retrospective chart review of all urine specimens in a 2-year period that screened positive on Abbott Architect and Aeroset platform but tested negative during confirmation. The false-positive urine drug screen results were cross-referenced with prescription drugs these patients received. The authors observed that 26.5% urine specimens had false-positive benzodiazepine results due to sertraline use.[49]

Efavirenz, a nonnucleoside reversetranscriptase inhibitor, is used for treating patients infected with human immunodeficiency virus (HIV). However, patient taking this drug may be tested false positive for benzodiazepine using the Biosite Triage 8 point-of-care testing kit. Blank et al. tested urine samples obtained from 100 patients infected with HIV. Of the 50 patients who received efavirenz, 49 patients with proven efavirenz intake tested positive for benzodiazepines with the Triage 8. However, of the 50 patients who did not receive efavirenz, only one tested positive for benzodiazepines. In the efavirenz group, efavirenz concentrations (as determined by LC-MS/MS) were below the limit of quantification. Measurement of metabolites showed mean 8-hydroxy-efavirenz concentrations of 238.4 μg/mL (range, 33.7–678.7 μg/mL) and mean dihydroxy-efavirenz concentrations of 36.9 μg/mL (range, 3.7–69.8 μg/mL). Therefore the major metabolite 8-hydroxy-efavirenz is responsible for this cross-reactivity. In addition, in the efavirenz group, 49 patients also tested positive for tetrahydrocannabinol with the Triage 8 and 46 patients tested positive in the drug screen using Multi 5.[50]

For determination of compliance of patients with benzodiazepine therapy, often drug testing is conducted in urine using 200 ng/mL cutoff, but immunoassays for benzodiazepines lack the requisite sensitivity for detecting benzodiazepine use in this population primarily due to their poor cross-reactivity with several major urinary benzodiazepine metabolites. Darragh evaluated the sensitivity of CEDIA in benzodiazepine detection (a high sensitivity cloned enzyme donor immunoassay (HS-CEDIA), in which β-glucuronidase is added to the reagent) and KIMS screening immunoassays for benzodiazepines by comparing results obtained by these immunoassays with LC-MS/MS. A total of 299 urine specimens from patients treated for chronic pain were analyzed and 141 (47%) confirmed positive for one or more of the benzodiazepines/metabolites by LC-MS/MS. However, CEDIA tested positive for 78 out of 141 confirmed-positive specimens, KIMS tested positive for 66 specimens, and HS-CEDIA correctly identified 110 specimens, primarily due to increased detection of lorazepam, but HS-CEDIA still missed 22% (31/141) of benzodiazepine-positive urine specimens. The authors concluded that although HS-CEDIA provides higher sensitivity than KIMS and CEDIA, it still missed an unacceptably high percentage of benzodiazepine-positive samples from patients treated for chronic pain.[51]

INTERFERENCE IN OPIATE IMMUNOASSAYS

In order to circumvent false-positive test results due to ingestion of poppy seed–containing food in opiate assays, the current SAMHSA guideline recommends a cutoff of 2000 ng/mL for opiate screening tests.

However, private employers may still use the old cut-off concentration of 300 ng/mL for opiate, and consumption of poppy seed–containing food can easily result in a positive screening as well as confirmation of codeine and morphine by GC/MS. The antibody in the opiate immunoassay targets morphine but can detect codeine. However, many opioids such as oxycodone- and oxymorphone have poor cross-reactivity with opiate immunoassays. These drugs may not be detected by opiate immunoassays. Smith et al. commented that, in general, immunoassays for opiates displayed substantially lower sensitivity for detecting 6-keto opioids, and urine specimens containing low to moderate amounts of hydromorphone, hydrocodone, oxymorphone, and oxycodone will likely be negative if analyzed using opiate immunoassays.[52] In addition, many opioids including buprenorphine, fentanyl, methadone, meperidine, propoxyphene, tramadol, etc. are not detected by opiate assays. Therefore opiate immunoassays are not suitable for detecting the presence of these drugs in urine. Sometimes urine drug tests are conducted in patients taking oxycodone and related drugs to determine compliance. However, opiate immunoassays are inappropriate for drug testing in these patients. Therefore specific immunoassays such as DRI oxycodone, DRI hydrocodone/hydromorphone, and methadone, buprenorphine, propoxyphene,fentanyl, and tramadol immunoassays must be used for monitoring the presence of these drugs in urine.

Case Report

A 40-year-old man receiving 20-mg oxycodone (Oxy-Contin) twice a day routinely for headache called the clinic stating that he finished his medication faster and needed a refill. His urine drug test result was negative using opiate immunoassay and suspecting he was selling oxycodone tablets, he was dismissed from the clinic. A family member contacted a toxicologist who informed that oxycodone may cause false-negative urine opiate drug screen results. An aliquot of the original urine specimen was retested using GC/MS and oxycodone level of 1124 ng/mL was confirmed.[53]

Heroin is first metabolized to 6-acetylmorphine and then further transformed into morphine by a liver enzyme. Presence of 6-monoacetylmorphine (6-acetylmorphine) in urine is considered as the confirmation of heroin abuse. Although 6-acetylmorphine is further metabolized into morphine, which can be detected easily by a standard opiate immunoassay, there are also commercially available immunoassays for detection of 6-acetylmorphine in urine at a cutoff concentration of 10 ng/mL.

Opiate immunoassays like other immunoassays also provide false-positive test results due to presence of various cross-reacting substances other than opioid in urine specimens. Certain quinolone antibiotics may cause false-positive test results with opiate immunoassay screening. Baden evaluated potential interference of 13 commonly used quinolones (levofloxacin, ofloxacin, pefloxacin, enoxacin, moxifloxacin, gatifloxacin, trovafloxacin, sparfloxacin, lomefloxacin, ciprofloxacin, clinafloxacin, norfloxacin, and nalidixic acid) with various opiate immunoassays (at 300 ng/mL cutoff concentration) and observed that levofloxacin and ofloxacin may cause false-positive opiate test results with assays manufactured by Abbot Laboratories for application on the AxSYM analyzer (Abbott Laboratories). In addition, such interferences were also observed with CEDIA, EMIT II, and Abuscreen ONLINE assay (Roche Diagnostics). Moreover, pefloxacin administration may produce false-positive results with CEDIA, EMIT II, and Abuscreen ONLINE assay; gatifloxacin administration with CEDIA and EMIT II assays;and lomefloxacin, moxifloxacin, ciprofloxacin, and norfloxacin with the Abuscreen ONLINE assay.[54] Rifampicin is used in treating tuberculosis and may cause false-positive test results with opiate immunoassays such as the KIMS assay on the cobas Integra analyzer (Roche Diagnostics). A false-positive result may be observed even after 18 h of administration of a single oral dose of 600 mg of rifampicin.[55]

Pholcodine, an opioid cough suppressant (antitussive), used in suppressing unproductive coughs is known to interfere with opiate immunoassays. Morphine is a minor metabolite of pholcodine.[56] Naloxone (Narcan), an opioid antagonist, also interferes with opiate immunoassays.

Case Report

A 3-year-old girl was transferred to the emergency department after being found shivering and unattended outside a public shopping area in winter. She presented with decreased mental status and possible hypothermia. Her physical examination revealed multiple abrasions and bruises. The patient had a temperature of 37°C, a blood pressure of 134/74 mmHg, a heart rate of 103 beats/min, and a respiratory rate of 24 breaths/min. Computed tomography scans, a skeletal survey, and an ophthalmologic examination showed no injury. An intravenous bolus of the opioid antagonist naloxone (Narcan) was administered shortly after her arrival in the emergency department. Routine hematologic and chemical tests were performed and revealed decreased serum urea nitrogen, along with increased lactate, aspartate aminotransferase, and alanine aminotransferase levels. Her urine drug screen result was negative and only acetone was detected in serum by GC. She was admitted to a pediatric floor for further monitoring. Early the next day, clinicians requested a second urine toxicologic immunoassay screen, which was positive for opiate using the CEDIA at 300 ng/mL cutoff. However, no opiate/opioid was confirmed by GC/MS. Suspecting false-positive opiate screening test result due to naloxone cross-reactivity with assay antibody, the authors performed in vitro experiments and observed positive opiate response when drug-free urine was supplemented with 6100 ng/mL naloxone. Further analysis of the second urine sample, reported as "opiate positive," showed a naloxone level of 9900 ng/mL explaining the false-positive test result.[57]

Interference in Methadone Immunoassays

False-positive test results with methadone immunoassays due to the presence of interfering substances in urine have been reported. In one report the authors observed false-positive methadone test results using the cobas Integra Methadone II test kit (Roche Diagnostics) in three schizophrenic patients treated with quetiapine monotherapy. The authors used a 300-ng/mL cutoff concentration of methadone for screening of urine specimens. However, using LC-MS, no methadone was detected in the plasma specimen of any patient.[58] In another report, urine from a patient tested positive for methadone (using cobas Integra Methadone II test kit based on KIMS methodology and manufactured by Roche Diagnostics) without a history of methadone ingestion. Drugs that have been shown to cross-react with methadone feature a tricyclic structure with a sulfur and nitrogen atom in the middle ring, which is common for both quetiapine and methadone. Therefore it is plausible that this structural similarity between quetiapine and methadone could underlie the cross-reactivity in methadone drug screening. Quetiapine at a dosage of 125 mg/day is sufficient to cause a false-positive methadone test result. The authors recommended verification of the test results with a different screening test or additional analytic tests in order to avoid adverse consequences for the patients.[59] Antipsychotic drugs chlorpromazine and thioridazine, as well as the antidepressant clomipramine, are also known to cross-react with methadone immunoassays.[35]

Rogers et al. reported positive methadone urine drug test results in a patient using the One Step Multi-Drug, Multi-Line Screen Test Devices (ACON Laboratories, San Diego, CA), a point-of-care device for urine drug screen. The patient had no history of methadone exposure but ingested diphenhydramine. The GC/MS confirmatory test failed to detect the presence of methadone in the urine specimen, confirming that the presumptive methadone test result was a false-positive result. When drug-free urine specimens were supplemented with diphenhydramine, false-positive methadone test results were also observed using the point-of-care device.[60] Doxylamine intoxication may cause false-positive results with both EMIT d.a.u opiate and methadone assays. The urine doxylamine concentration needed to cause positive test result was 50 μg/mL for methadone and 800 μg/mL for opiate.[61] Verapamil metabolites also interfere with methadone assay for application on Olympus AU5000 analyzer.[62]

Tapentadol, a synthetic analgesic, interferes with the DRI methadone assay (Thermo Fisher) at 130 ng/mL cutoff. In one study, 97 authentic tapentadol urine specimens that produced false-positive methadone results were analyzed for methadone and tapentadol in compound-specific ultraperformance liquid chromatography-MS-MS confirmation tests. Tapentadol, tapentadol glucuronide, tapentadol sulfate, and N-desmethyltapentadol exhibited cross-reactivity with the methadone enzyme immunoassay (EIA) at 6500 (2.2%), 25,000 (0.6%), 3000 (4.4%), and 20,000 ng/mL (0.9%) concentrations, respectively. No cross-reactivity was observed with the methadone metabolite 2-ethylidine-1,5-dimethyl-3,3-diphenylpyrrolidine using EIA. All authentic urine specimens were confirmed to be negative for methadone but positive for tapentadol and all monitored metabolites, indicating that taking tapentadol may cause false-positive methadone results using the DRI methadone assay.[63] However, in another study the authors observed no false-positive test results using the Syva EMIT II methadone assay in all 11 patients reported who were taking therapeutic dosages of tapentadol between 50 and 600 mg/day.[64]

Interferences in Immunoassays for Other Opioids

Buprenorphine CEDIA is subjected to cross-reactivity from tramadol, a synthetic opioid, at a cutoff of 5 ng/mL. However, when cutoff was increased to 20 ng/mL, this interference was eliminated.[65] Opiate, if present in high concentrations, may also cause false-positive buprenorphine immunoassay test results. In one study the authors reviewed data on falsely positive buprenorphine immunoassay screen results (cutoff ≥ 5 ng/mL) but negative for buprenorphine GC/MS in urine specimens, which also tested positive by opiate immunoassay at 300 ng/mL cutoff. The authors found that cross-reactivity in the CEDIA buprenorphine immunoassay by opiates at concentrations <2000 ng/mL will not cause a false-positive buprenorphine result.[66] In addition, morphine, codeine, and dihydrocodeine, as well as amisulpride or sulpiride, if present in high concentrations in urine are known to interfere with the CEDIA buprenorphine assay.[67]

Case Report

A 57-year-old man who contracted vertebral osteomyelitis 15 years ago presented to the hospital with complicated clinical history. He underwent multiple orthopedic procedures to stabilize his lumbar spine.

After years of titration, he was comfortable on his current analgesic regimen of 200 mg of pregabalin taken orally 3 times per day as well as 360 mg of oxycodone and 960 mg of morphine per day. In monitoring this patient, the authors noted consistent positive results for buprenorphine. The patient was not taking buprenorphine, and GC/MS confirmation was negative indicating false-positive test results using immunoassay. The authors discontinued oxycodone for a period and then discontinued morphine. When oxycodone was discontinued, his urine test results were positive for buprenorphine using immunoassay but negative by GC/MS confirmation. However, when morphine was discontinued, the urine specimen tested negative by buprenorphine immunoassay, indicating that false-positive test result was due to morphine. Further investigation revealed that when morphine levels in urine exceeded 15,000 ng/mL, false-positive buprenorphine immunoassay test results were observed.[68]

Fentanyl and its analogs, such as acetyl-fentanyl, have become a concern for potential abuse. The DRI fentanyl immunoassay showed cross-reactivity from acetyl-fentanyl, risperidone, and its metabolite 9-hydroxyrisperidone.[69]

INTERFERENCES IN MARIJUANA IMMUNOASSAYS

The active component of marijuana, Δ^9-tetrahydrocannabinol, is metabolized into THC-COOH. Then THC-COOH is conjugated with glucuronic acid. The immunoassays screening for marijuana target THC-COOH. Passive inhalation of marijuana should not cause false-positive test result with immunoassays because in one study the maximum concentration of THC-COOH was only 7.8 ng/mL after volunteers were exposed to passive inhalation of marijuana smoke.[70] Similarly use of hemp oil should not produce positive marijuana test results because hemp seeds are washed with water prior to extraction of oil, a procedure that removes traces of marijuana from the seed hull. However, prescription use of synthetic marijuana (Marinol) should cause positive marijuana test results.

False-positive immunoassay screening results for marijuana have been reported. Boucher et al. described the case of a 3-year-old girl who was hospitalized because of behavioral disturbance of unknown cause. The only remarkable finding in her medication history was the suppositories of niflumic acid, which was initiated 5 days before hospitalization. After admission, her urinary toxicologic screen result was positive for the presence of marijuana metabolite but parents strongly denied such exposure. Further analysis of the specimen using chromatography failed to confirm the presence of marijuana metabolite but niflumic acid was detected in the specimen. The authors concluded that the false-positive marijuana test result was due to the presence of niflumic acid in the urine specimen.[71] The antiviral agent efavirenz is known to cross-react with marijuana immunoassays. In one study the authors analyzed 30 urine specimens collected from patients receiving efavirenz by using the Rapid Response drugs of abuse test strips, the Beckman Coulter SYNCHRON marijuana immunoassay (Beckman Coulter, Brea, CA), and the Roche Diagnostics Cannabinoid II assay. Only the Rapid Response test strips demonstrated positive marijuana test results in 28 out of 30 specimens, whereas the two other immunoassays did not show any interference from efavirenz. As expected, GC/MS confirmation failed to demonstrate the presence of marijuana metabolite in any of the 30 specimens analyzed.[72] The nonsteroidal drugs naproxen and ibuprofen may cause false-positive test results in marijuana immunoassay.[73]

INTERFERENCES IN PHENCYCLIDINE IMMUNOASSAYS

False-positive test results have been reported in PCP immunoassays due to cross-reactivity of several drugs. Dextromethorphan is an antitussive agent that is found in many over–the-counter cough and cold medications. Ingesting high amounts of dextromethorphan (over 30 mg) may result in false-positive test results with opiate and PCP immunoassays. In one report the authors observed three false-positive PCP test results in pediatric urine specimens using an on-site testing device (Instant-View Multi-Test drugs of abuse panel; Alfa Scientific Designs, Poway, CA). The authors concluded that false-positive PCP test results were due to the cross-reactivities of ibuprofen, metamizole, dextromethorphan, and their metabolites with the PCP assay.[74] Thioridazine is known to cause false-positive PCP test results with both EMIT d.a.u and EMIT II PCP immunoassays.[75] Venlafaxine and its metabolites also may cause false-positive PCP test results in immunoassay.[76] Ketamine administration can produce a false-positive urine PCP screen result using immunoassay.[77]

Tramadol when present in very high concentrations may cause false-positive test results with the DRI PCP assay (Thermo Fisher). Ly et al. presented two cases (one case where a 31-year-old man ingested 84 tramadol tablets each containing 50 mg tramadol in a suicide attempt and another case where the amount of tramadol taken was unknown) of false-positive PCP qualitative urine drug screen results due to tramadol misuse. To verify that tramadol or one of its metabolites (O-desmethyltramadol, N-desmethyltramadol) were responsible for triggering such false-positive PCP immunoassay results, the authors supplemented drug-free urine with pure standards and observed that tramadol and N-desmethyltramadol at high concentrations of 500, 000 ng/mL produced results above the positive cutoff (25 ng/mL) for the PCP immunoassay.[78] In another study the authors reported two patients with positive PCP screen results using the Emit II Plus PCP assay (Siemens) after tramadol ingestion.[79]

INTERFERENCES IN MISCELLANEOUS OTHER IMMUNOASSAYS

Barbiturate immunoassays can recognize a wide variety of barbiturates including amobarbital, butalbital, pentobarbital, secobarbital, and phenobarbital. Secobarbital is used as a calibrator in several commercially available immunoassays for barbiturates. Acute ingestion of ibuprofen and chronic ingestion of naproxen may cause false-positive test results with barbiturate immunoassay.[73] False-positive ketamine immunoassay test results may occur due to ingestion of quetiapine.

Case Report

A 33-year-old male was admitted to acute psychiatric ward due to mania-like symptoms for 4 months. He had a history of amphetamine use since 5 years ago, and quitted for 2 years. His initial urine drug screen result was negative. However, when diazepam was discontinued and quetiapine 25 mg was added on the 12th day, and quetiapine dose was titrated to 50 mg on the 15th day, the patient showed positive ketamine urine drug screen on the 21st day. The specimen was sent to a reference laboratory, where it again tested positive by the DRI ketamine assay (Thermo Fisher) but negative by GC/MS confirmation. The authors concluded that quetiapine may cause false-positive ketamine immunoassay test results.[80]

Administration of mucolytic agent ambroxol may cause false-positive test results with immunoassays used for screening LSD in urine. In one study the authors reported that urine specimens collected from 10 patients

TABLE 7.5

Interferences in Immunoassays for Barbiturates, Benzodiazepines, Bupropion, Ketamine, Opiate, Fentanyl, Methadone, Marijuana, and Phencyclidine

Immunoassay	Interfering Drugs
Barbiturates	Ibuprofen and naproxen
Benzodiazepines	Oxaprozin, efavirenz, sertraline
Bupropion	Morphine, tramadol, amisulpride, sulpiride
Ketamine	Quetiapine
Opiate	Antibiotic (levofloxacin, ofloxacin, pefloxacin, gatifloxacin, lomefloxacin, moxifloxacin, ciprofloxacin, and norfloxacin), pholcodine, doxylamine
Fentanyl	Acetyl-fentanyl, risperidone, and its metabolite 9-hydroxyrisperidone
Methadone	Clomipramine, chlorpromazine, thioridazine, quetiapine, diphenhydramine, doxylamine, and verapamil
Marijuana	Niflumic acid, efavirenz, ibuprofen, and naproxen
Phencyclidine	Dextromethorphan, ibuprofen, metamizole, thioridazine, tramadol (severe overdose), venlafaxine, ketamine
Lysergic acid diethylamide	Ambroxol, sertraline

receiving ambroxol showed positive LSD test result using CEDIA for LSD, but the presence of LSD could not be confirmed by the HPLC technique. However, the presence of ambroxol was confirmed in urine by HPLC.[81] Röhrich et al. also reported unexpected positive results for LSD in urine samples from 12 patients in an intensive care unit in a routine screening using the CEDIA drugs of abuse assay. Ketamine could not be confirmed by HPLC in these specimens, but they all contained the mucolytic drug ambroxol. Further studies demonstrated that ambroxol exhibits a significant cross-reactivity in the CEDIA for LSD.[82] Sertraline also interferes with the CEDIA for LSD if present in a concentration of 1.5 μg/mL or more in urine.[83] Interferences in immunoassays for barbiturates, benzodiazepines, bupropion, ketamine, opiate, fentanyl, methadone, marijuana, and PCP are summarized in Table 7.5.

EFFECT OF URINE ADULTERATION ON DRUGS OF ABUSE TESTING

People try to beat drug test in different ways:

- Drinking detoxifying agent with a hope to excrete illicit drugs from the body
- Substituting urine specimen by a drug-free specimen
- Submitting synthetic urine instead of one's own urine
- Diluting urine with tap water
- Adding in vitro adulterants to beat drug test

The purpose of beating drug test is to produce false-negative test results. However, drug testing laboratories routinely perform "specimen validity testing" to identify adulterated urine specimens before analysis. At the collection site the temperature of the urine specimen must be determined and the temperature should be between 90 and 100 °F. This step may be helpful to identify substituted or diluted urine because many collection facilities do not have hot water in the bathroom for specimen collection. According to the updated SAMSHA guidelines published in 2008, all laboratory involved in federal workplace drug testing must perform analysis of creatinine, analyze specific gravity if creatinine is less than 20 mg/dL, and measure the pH of the specimen. In addition, tests for the presence of oxidants, and more tests if needed, must be conducted if the specimen is suspected to contain any adulterant.[2]

Various adulterants are used to maskdrugs. However, when household chemicals such as table salt (sodium chloride), vinegar, laundry bleach, hand soap, Drano, sodium bicarbonate, sodium hypochlorite, and concentrated lemon juice are used, such adulterated specimens can be easily identified through the specimen validity testing (creatinine, pH, and specific gravity). In addition, Visine eye drop is capable of producing false-negative test results with marijuana during immunoassay screening (no effect on GC/MS confirmation) but the presence of Visine eye drop in adulterated urine cannot be determined by urine specimen integrity testing.[84]

Many urinary adulterants are available through the Internet but they may contain the same active ingredient. Wu et al. reported that the active ingredient of Urine Luck was pyridinium chlorochromate (PCC), a strong oxidizing agent that caused significantly decreased response rate for all EMIT II drug screens producing false-negative results. However, Wu et al.[85] also described a simple spot test using 1,5-diphenylcarbazide in methanol (10 g/L) to detect the presence of PCC in urine (reddish purple color developed in the presence of PCC). Other adulterants available through the Internet such as Clear Choice, Lucky Lab LL 418, Randy's Klear II, and Sweet Pea's Spoiler also contain PCC and have similar effects like Urine Luck on drugs of abuse testing. The DRI General Oxidant-Detect test can detect the presence of PCC in urine. In addition, special urine dipsticks designed for detecting urinary adulterants have specific reaction pads for detecting PCC.

The product "Klear" is supplied in the form of two small tubes containing 500 mg of white crystalline material that ElSohly et al. identified as potassium nitrite and also provided evidence that nitrite adulteration caused interference in both immunoassay screening and GC/MS confirmation of THC-COOH. However, a bisulfite step at the beginning of sample preparation could eliminate such problems.[86] Spot tests are available for detecting the presence of PCC or nitrite in urine.[87] Specialized urine dipsticks have specific pads for detecting nitrite. The DRI General Oxidant-Detect test is also capable of detecting nitrite if present in the adulterated urine. However, a different analytic method must be used for confirmation.

Stealth is an adulterant that consists of two vials, one containing a powder (peroxidase) and another vial containing a liquid (hydrogen peroxide). Both products should be added to the urine specimen in order to cheat drug test. Stealth is capable of invalidating immunoassay screening of marijuana, LSD, and opiates using both Roche ONLINE assays and CEDIA if these drug or metabolites are present in modest concentrations (125%–150% of cutoff values). In addition, GC/MS confirmation could also be affected.[88] Valtier and Cody[89] described a rapid spot test to detect the presence of Stealth in urine. If Stealth is present, the DRI General Oxidant-Detect test should be able to detect its presence. Moreover, urine dipsticks can be used for detecting the presence of Stealth in urine.

Glutaraldehyde-containing products such as "Urinaid" are one of the first products to appear in the market to invalidate drugs of abuse testing. Glutaraldehyde at a concentration of 0.75% volume could lead to false-negative screening results for a cannabinoid test using the EMIT II drugs of abuse screen. At 2% glutaraldehyde concentration, almost all EMIT II assays were affected. Wu et al.[90] also described a simple fluorometric method for the detection of glutaraldehyde in urine. Specialized urine dipsticks have a pad for detecting glutaraldehyde.

SAMHSA published extensive guidelines for detecting the presence of any urine adulterants in the specimens prior to immunoassay screening and a further confirmation step if needed because many adulterants may cause false-negative test results. Various analytic techniques such as colorimetry using multiwavelength spectrophotometry, capillary electrophoresis or ion chromatography capillary electrophoresis, or inductively coupled plasma mass spectrometry can be used for screening and confirming the presence of any urinary adulterant. The presence of surfactant is verified by using a surfactant colorimetric test with ≥100 µg/mL dodecylbenzene sulfonate equivalent and confirmation at the same level by using a different analytic method, for example, multiwavelength spectrophotometry. Detailed discussion on this topic is beyond the scope of this book, but Fu[91] has reviewed this topic.

ISSUES OF INTERFERENCE IN TOXICOLOGY TESTING

Although serum level of alcohol is measured using an enzymatic assay, not an immunoassay, it is an important analyte that is analyzed in toxicology laboratories and there are issues of interferences in enzymatic alcohol assays that utilize conversion of alcohol to acetaldehyde by alcohol dehydrogenase enzyme,where NAD is converted into NADH. While NADH has no absorption at 340 nm, NADH absorbs at 340 nm and the signal intensity is proportional to the alcohol concentration in the specimen. High amount of lactate and lactate dehydrogenase (LDH) if present in the specimen interferes with this assay causing false-positive test results because LDH converts lactate into pyruvate and in this process NAD is also converted into NADH. High lactate and LDH concentrations are observed in postmortem blood. Therefore enzymatic alcohol assays are unsuitable for determination of blood alcohol in postmortem blood. Moreover, any legal alcohol testing must be conducted using headspace GC.

Lactate concentrations tend to increase in trauma patients. Dunne et al.[92] reported that 27% (3536) of the 13,102 patients they studied had positive alcohol screen (mean alcohol, 141 mg/dL; range, 10–508 mg/dL). Alcohol is not protein bound in serum. Therefore interference of lactate and LDH in enzymatic alcohol assay can also be eliminated by measuring free alcohol (alcohol concentration in protein-free ultrafiltrate) using enzymatic alcohol assay because LDH (molecular weight, 140 kDa) is absent in the protein-free ultrafiltrate.[93]

Because the legal limit of alcohol while driving is 0.08% (80 mg/dL) of whole blood, if serum alcohol is used as evidence in a court of law (medical alcohol is sometimes accepted in a court of law as evidence for criminal trial of alcohol-related driving and other incidences), it must be converted into whole blood alcohol. Rainey reported that the ratio between serum and whole blood alcohol ranged from 0.88 to 1.59 but the median

value was 1.15. Therefore dividing serum alcohol value by 1.15 provides whole blood alcohol concentration, or multiplying serum alcohol by 0.87 should also provide whole blood alcohol concentration.[94]

Acetaminophen and salicylate are sometimes used in suicide attempts. Serum level of acetaminophen is important in determining the dose of acetaminophen antidote Mucomyst (acetylcysteine) to be administered for treating severe acetaminophen overdose. Acetaminophen can be measured using enzymatic colorimetric assay or immunoassay. In enzymatic colorimetric assay, measured acetaminophen is hydrolyzed into p-aminophenol catalyzed by arylacylamidase. Then condensation of p-aminophenol with o-cresol in the presence of periodate forms the blue indophenol chromophore, which is measured at 600 nm wavelength. High bilirubin concentrations in serum may cause false-positive acetaminophen test results.

Case Report

A severely jaundiced 17-year-old male patient presented in the emergency department with abdominal pain. His liver enzyme as well as total bilirubin (19.8 mg/dL; reference interval, 3–1.2 mg/dL) concentrations were significantly elevated but serologic test result was negative for hepatitis A, B, and C. In addition, the patient also showed a serum acetaminophen concentration of 34 µg/mL. However, the patient denied using any medications containing acetaminophen within the previous week. The clinician contacted the laboratory to question the accuracy of the acetaminophen measurement, which was confirmed by reanalysis. Review of the package insert revealed that elevated bilirubin (25 mg/dL and above) may cause false-positive acetaminophen test results. Suspecting bilirubin interference, the authors selected 12 hyperbilirubinemic plasma specimens (total bilirubin, 15.9–33.8 mg/dL) from patients without a recent history of acetaminophen ingestion. Apparent acetaminophen concentrations of 5–18 µg/mL were detected in these specimens. Serial dilutions indicated that bilirubin interference was eliminated when bilirubin levels were <5 mg/dL.[95]

Polson et al.[96] commented that colorimetric enzymatic method for detection of acetaminophen is affected if bilirubin concentration in serum exceeds 10 mg/dL but immunoassays for acetaminophen (EMIT and FPIA) are not affected.

Significant positive bias of bilirubin in the colorimetric Trinder reaction based salicylate methods (color complex formed by reaction of salicylate with ferric ions) on automated analyzers has been reported. However, such interferences can be eliminated by using salicylate immunoassay.[97] Broughton[98] also described interference of bilirubin on a salicylate assay performed using the Olympus automated analyzer and Trinder method. Mitochondrial acetoacetyl-CoAthiolase deficiency is a rare metabolic disorder causing acute episodes of severe ketosis and acidosis. Tilbrook reported false-positive results for salicylate in an 18-month-old boy who was presented to the hospital with sever acidosis. The authors concluded that false-positive results for salicylate using Trinder reagent was due to interference of high levels of acetoacetate in the specimen.[99] However, immunoassays for salicylate manufactured by various diagnostic companies are free from interferences.

CONCLUSIONS

The first step in drugs of abuse testing is screening of urine specimens using FDA-approved immunoassays. For medical drug testing, confirmations using GC/MS or LC-MS/MS are usually not performed unless specifically requested by clinicians. In contrast, for legal drug testing, including workplace drug testing, confirmation step is mandatory. Immunoassays used for screening urine for the presence of drugs suffer from interferences. The immunoassays used for screening of amphetamine/methamphetamine are more affected by various drugs (most commonly due to the presence of ephedrine/pseudoephedrine in many over–the-counter cold medications) than other immunoassays, although several antibiotics may cause false-positive test results with opiate immunoassays if present in the specimen. However, confirmation methods using chromatography are robust and relatively free from any interference.

Enzymatic alcohol assay may be affected by high LDH and lactate concentrations but the gold standard for alcohol measurement, headspace GC, is free from interference. Moreover, whole blood can be analyzed for the presence of alcohol as well as other volatiles such as methyl alcohol, isopropyl alcohol, ethylene glycol, and acetone. Enzymatic acetaminophen assay and Trinder salicylate colorimetric assays are affected by high bilirubin levels but immunoassays are free from such interferences.

REFERENCES

1. Stout PR, Klette KL, Wiegand R. Comparison and evaluation of DRI methamphetamine, DRI ecstasy, Abuscreen ONLINE amphetamine, and a modified Abuscreen ONLINE amphetamine screening immunoassays for the detection of amphetamine (AMP), methamphetamine (MTH), 3,4-methylenedioxyamphetamine (MDA), and 3,4-methylenedioxymethamphetamine (MDMA) in human urine. *J Anal Toxicol.* 2003;27:265–269.
2. Bush D. The US mandatory guidelines for Federal workplace drug testing programs: current status and future considerations. *Forensic Sci Int.* 2008;174:111–119.
3. Department of Health and Human Services. Proposed revision to mandatory guidelines for federal workplace drug testing programs. *Fed Regist.* 2015;80:28101–28151.
4. Poklis A, Fitzgerald RL, Hall KV, Saddy JJ. Emit d.a.u monoclonal amphetamine/methamphetamine assay II. Detection of methylenedioxyamphetamine (MDA) and methylenedioxymethamphetamine (MDMA). *Forensic Sci Int.* 1993;59:63–70.
5. Lekskulchai V, Mokkhavesa C. Evaluation of Roche Abuscreen ONLINE amphetamine immunoassay for screening of new amphetamine analogs. *J Anal Toxicol.* 2001;25:471–475.
6. Kerrigan S, Mellon MB, Banuelos S, Arndt C. Evaluation of commercial enzyme-linked immunosorbent assays to identify psychedelic phenethylamine. *J Anal Toxicol.* 2011;25:444–451.
7. Apollonio LG, Whittall IR, Pianca DJ, Kyd JM, et al. Matrix effect and cross-reactivity of select amphetamine-type substances, designer analogues, and putrefactive amines using the Bio-Quant direct ELISA presumptive assays for amphetamines and methamphetamines. *J Anal Toxicol.* 2007;31:208–213.
8. Petrie M, Lynch KL, Ekins S, Chang JS, et al. Cross-reactivity studies and predictive modeling of "Bath salts" and other amphetamine type stimulants with amphetamine screening immunoassays. *Clin Toxicol.* 2013;51:83–91.
9. Ellefsen KN, Anizan S, Castaneto MS, Desrosiers NA, et al. Validation of the only commercially available immunoassay for synthetic cathinones in urine: Randox Drugs of Abuse V Biochip Array Technology. *Drug Test Anal.* 2014;6:728–738.
10. Poklis A, Jortani WSA, Brown CS, Crooks CR. Response of the EMIT II amphetamine/methamphetamine assay to specimens collected following use of Vicks inhalers. *J Anal Toxicol.* 1993;17:284–286.
11. Smith ML, Nichols DC, Underwood P, Fuller Z, et al. Methamphetamine and amphetamine isomer concentrations in human urine following controlled Vicks VapoInhaler administration. *J Anal Toxicol.* 2014;38:524–527.
12. Strano-Rossi S, Leone D, de la Torre X, Botrè F. The relevance of the urinary concentration of ephedrines in anti-doping analysis: determination of pseudoephedrine, cathine, and ephedrine after administration of over-the-counter medicaments. *Ther Drug Monit.* 2009;31:520–526.
13. DePriest AZ, Knight J, Doering PL, Black DL. Pseudoephedrine and false-positive immunoassay urine drug tests for amphetamine. *Pharmacotherapy.* 2013;33:e88–e89.
14. Dietzen DJ, Ecos K, Friedman D, Beason S. Positive predictive values of abused drug immunoassays on the Beckman SYNCHRON in a Veteran population. *J Anal Toxicol.* 2001;25:174–178.

15. Liu L, Wheeler SE, Rymer JA, Lower D, et al. Ranitidine interference with standard amphetamine immunoassay. *Clin Chim Acta.* 2015;438:307–308.

16. Casey ER, Scott MG, Tang S, Mullins ME. Frequency of false positive amphetamine screens due to bupropion using the Syva EMIT II immunoassay. *J Med Toxicol.* 2011;7:105–108.

17. Logan BK, Costantino AG, Rieders EF, Sanders D. Trazodone, meta-chlorophenylpiperazine (an hallucinogenic drug and trazodone metabolite), and the hallucinogen trifluoromethylphenylpiperazine cross-react with the EMIT®II ecstasy immunoassay in urine. *J Anal Toxicol.* 2010;34:587–589.

18. Baron JM, Griggs DA, Nixon AL, Long WH, et al. The trazodone metabolite meta-chlorophenylpiperazine can cause false positive urine amphetamine immunoassay result. *J Anal Toxicol.* 2011;35:364–368.

19. Gilbert RB, Peng PI, Wong D. A labetalol metabolite with analytical characteristics resembling amphetamines. *J Anal Toxicol.* 1995;19:84–86.

20. Yee LM, Wu D. False positive amphetamine toxicology screen results in three pregnant women using labetalol. *Obstet Gynecol.* 2011;117(2 Pt2):503–506.

21. Fucci N. False positive results for amphetamine in urine of a patient with diabetes mellitus. *Forensic Sci Int.* 2012;223:e60.

22. Papa P, Rocchi L, Mainardi C, Donzelli G. Buflomedil interference with the monoclonal EMIT d.a.u. amphetamine/methamphetamine immunoassay. *Eur J Clin Chem Clin Biochem.* 1997;35:369–370.

23. Snozek CLH, Kaleta EJ, Jannetto PJ, Linkenmeyer JJ, et al. False-positive amphetamine results on several drug screening platforms due to mexiletine. *Clin Biochem.* 2008;58:125–127.

24. Gomila I, Quesada L, López-Corominas V, Fernández J, et al. Cross-reactivity of chloroquine and hydroxychloroquine with DRI amphetamine immunoassay. *Ther Drug Monit.* 2017;39:192–196.

25. Lora-Tamayo C, Tena T, Rodríguez A, Moreno D. High concentration of chloroquine in urine gives positive result with Amphetamine CEDIA reagent. *J Anal Toxicol.* 2002;26:58.

26. Kaplan J, Shah P, Faley B, Siegel ME. Case reports of aripiprazole causing false-positive urine amphetamine drug screens in children. *Pediatrics.* 2015;136:e1625–1628.

27. Merigian KS, Beowning RG. Desipramine and amantadine causing false positive urine test for amphetamine. *Ann Emerg Med.* 1993;22:1927–1928.

28. Merigian KS, Browning R, Kellerman A. Doxepin causing false-positive urine test for amphetamine. *Ann Emerg Med.* 1993;22:1370.

29. Schmolke M, Hallbach J, Guder WG. False-positive results for urine amphetamine and opiate immunoassays in a patient intoxicated with perazine. *Clin Chem.* 1996;42(10):1725–1726.

30. Levine BS, Caplan YH. Isometheptene cross reacts in the EMIT amphetamine assay. *Clin Chem.* 1987;33:1264–1265.

31. Manzi S, Law T, Shannon MW. Methylphenidate produces a false-positive urine amphetamine screen. *Pediatr Emerg Care.* 2002;18:401.

32. Breindahl T, Hindersson P. Methylphenidate is distinguished from amphetamine in drug of abuse testing. *J Anal Toxicol.* 2012;36:538–539.

33. Vorce SP, Holler JM, Cawrse BM, Magluilo J. Dimethylamine: a drug causing positive immunoassay results for amphetamines. *J Anal Toxicol.* 2011;35:183–187.

34. Pavletic AJ, Pao M. Popular dietary supplement causes false-positive drug screen for amphetamines. *Psychosomatics.* 2014;55:206–207.

35. Brahm NC, Yeager LL, Fox MD, Farmer KC, et al. Commonly prescribed medications and potential false-positive urine drug screens. *Am J Health Syst Pharm.* 2010;67:1344–1350.

36. Kaplan YC, Erol A, Karadas B. False-positive amphetamine/ecstasy (MDMA/3,4-methylenedioxymethamphetamine) (CEDIA) and ecstasy (MDMA/3,4-methylenedioxymethamphetamine) (DRI) test results with fenofibrate. *Ther Drug Monit.* 2012;34:493–495.

37. Cody JT, Valtier S. Amphetamine, clobenzorex, and 4-hydroxyclobenzorex levels following multidose administration of clobenzorex. *J Anal Toxicol.* 2001;25:158–165.

38. Critchfield GC, Wilkins DG, Loughmiller DL, Davis BW, et al. Antibody-mediated interference of a homogeneous immunoassay. *J Anal Toxicol.* 1993;17:69–72.

39. Patrinely JR, Cruz OA, Reyna GS, King JW. The use of cocaine as an anesthetic in lacrimal surgery. *J Anal Toxicol.* 1994;18(1):54–56.

40. Jenkins AJ, Llosa T, Montoya I, Cone EJ. Identification and quantitation of alkaloids in coca tea. *Forensic Sci Int.* 1996;77:179–189.

41. Jackson GF, Saady JJ, Poklis A. Urinary excretion of benzoylecgonine following ingestion of health Inca tea. *Forensic Sci Int.* 1991;49:57–64.

42. Mazor SS, Mycyk MB, Wills B, Brace LD, et al. Coca tea consumption causes positive urine cocaine assay. *Eur J Emerg Med.* 2006;13:340–341.

43. Turner M, McCrory P, Johnston A. Time for tea anyone? *Br J Sports Med.* 2005;39:e37.

44. Baker JE, Jenkins AJ. Screening for cocaine metabolite fails to detect an intoxication. *Am J Forensic Med Pathol.* 2008;29:141–144.

45. Fraser AD, Howell P. Oxaprozin cross-reactivity in three commercial immunoassays for benzodiazepines in urine. *J Anal Toxicol.* 1998;22:50–54.

46. Reid KS. Positive interference of the analgesic nefopam in the urine immunoassay for benzodiazepines in a secure setting. *J Psychopharmacol.* 2009;23:997.

47. El-Haj B, Al-Amri A, Ali H. Cross-reactivity of nefopam and its metabolites with benzodiazepine EMIT immunoassay. *J Anal Toxicol.* 2008;32:790–792.

48. Fitzgerald RL, Herold DA. Improved CEDIA benzodiazepine assay eliminates sertraline crossreactivity. *J Anal Toxicol.* 1997;21:32–35.

49. Nasky KM, Cowan GL, Knittel DR. False-positive urine screening for benzodiazepines: an association with sertraline? A two-year retrospective chart analysis. *Psychiatry (Edgmont)*. 2009;6:36–39.

50. Blank A, Hellstern V, Schuster D, Hartmann M, et al. Efavirenz treatment and false-positive results in benzodiazepine screening tests. *Clin Infect Dis*. 2009;48:1787–1789.

51. Darragh A, Snyder ML, Ptolemy AS, Melanson SKIMS. CEDIA, and HS-CEDIA immunoassays are inadequately sensitive for detection of benzodiazepines in urine from patients treated for chronic pain. *Pain Physician*. 2014;17:359–366.

52. Smith ML, Hughes RO, Levine B, Dickerson S, et al. Forensic drug testing for opiates. VI. Urine testing for hydromorphone, hydrocodone, oxymorphone, and oxycodone with commercial opiate immunoassays and gas chromatography-mass spectrometry. *J Anal Toxicol*. 1995;19:18–26.

53. Von Seggern RL, Fitzerald CP, Adelman LC, Adelman JU. Laboratory monitoring of OxyContin (oxycodone): clinical pitfalls. *Headache*. 2004;44:44–47.

54. Baden LR, Horowitz G, Jacoby H, Eliopoulos GM. Quinolones and false positive urine screening for opiates by immunoassay technology. *J Am Med Assoc*. 2001;286:3115–3119.

55. De Paula M, Saiz LC, Gonzalez-Revalderia J, Pascual T, et al. Rifampicin causes false positive immunoassay results for opiates. *Clin Chem Lab Med*. 1998;36:241–243.

56. Johansen M, Rasmussen KE, Christophersen AS, Skuterud B. Metabolic study of pholcodine in urine using enzyme multiplied immunoassay technique (EMIT) and capillary gas chromatography. *Acta Pharm Nord*. 1991;3:91–94.

57. Straseski JA, Stolbach A, Clarke W. Opiate-positive immunoassay screen in a pediatric patient. *Clin Chem*. 2010;56:1220–1223.

58. Widschwendter CG, Zernig G, Hofer A. Quetiapine cross-reactivity with urine methadone immunoassays. *Am J Psychiatry*. 2007;164:172.

59. Lasić D, Uglesić B, Zuljan-Cvitanović M, Supe-Domić D, et al. False-positive methadone urine drug screen in a patient treated with quetiapine. *Acta Clin Croat*. 2012;51:269–272.

60. Rogers SC, Pruitt CW, Crouch DJ, Caravati EM. Rapid urine drug screens: diphenhydramine and methadone cross-reactivity. *Pediatr Emerg Care*. 2010;26:665–666.

61. Hausmann E, Kohl B, von Boehmer H, Wellhoner HH. False positive EMIT indication for opiates and methadone in doxylamine intoxication. *J Clin Chem Clin Biochem*. 1983;21:599–600.

62. Lichtenwalner MR, Mencken T, Tully R, Petosa M. False positive immunochemical screen for methadone attributable to metabolites of verapamil. *Clin Chem*. 1998;44:1039–1041.

63. Collins AA, Merritt AP, Bourland JA. Cross-reactivity of tapentadol specimens with DRI methadone enzyme immunoassay. *J Anal Toxicol*. 2012;36:582–587.

64. Mullins ME, Hock K, Scott MG. Does therapeutic use of tapentadol cause false-positive urine screens for methadone or opiates? *Clin Toxicol*. 2015;53:493–494.

65. Shaikh S, Hull MJ, Bishop KA, Griggs DA, et al. Effect of tramadol use on three point of care and one instrument based immunoassays for buprenorphine. *J Anal Toxicol*. 2008;32:339–343.

66. Saleem M, Martin H, Tolya A, Coates P. Do all screening immunoassay positive buprenorphine samples need to be confirmed? *Ann Clin Biochem*. 2017;54:707–711.

67. Birch MA, Couchman L, Pietromartire S, Karna T, et al. False-positive buprenorphine by CEDIA in patients prescribed amisulpride or sulpiride. *J Anal Toxicol*. 2013;37:233–236.

68. Tenore PL. False-positive buprenorphine EIA urine toxicology results due to high dose morphine: a case report. *J Addict Dis*. 2012;31:329–331.

69. Wang BT, Colby JM, Wu AH, Lynch KL. Cross-reactivity of acetylfentanyl and risperidone with a fentanyl immunoassay. *J Anal Toxicol*. 2014;38:672–675.

70. Rohrich J, Schimmel I, Zorntlein S, Becker J, et al. Concentrations of delta-9-tetrahydrocannabinol and 11-nor 9-carboxytetrahydrocannabinol in blood and urine after passive exposure to cannabis smoke in a coffee shop. *J Anal Toxicol*. 2010;34:196–203.

71. Boucher A, Vilette P, Crassard N, Bernard N, et al. Urinary toxicological screening: analytical interference between niflumic acid and cannabis. *Arch Pediatr*. 2009;16:1457–1460. [Article in French].

72. Oosthuizen NM, Laurens JB. Efavirenz interference in urine screening immunoassays for tetrahydrocannabinol. *Ann Clin Biochem*. 2012;49:194–196.

73. Rollins DE, Jennison TA, Jones G. Investigation of interference by nonsteroidal anti-inflammatory drugs in urine tests for abused drugs. *Clin Chem*. 1998;36:602–606.

74. Marchei E, Pellegrini M, Pichini S, Martin I, et al. Are false positive phencyclidine immunoassay instant-view multi test results caused by overdose concentrations of ibuprofen, metamizol and dextromethorphan? *Ther Drug Monit*. 2007;29:671–673.

75. Long C, Crifasi J, Maginn D. Interference of thioridazine (Mellaril) in identification of phencyclidine. *Clin Chem*. 1996;42:1885–1886.

76. Bond GR, Steele PE, Uges DR. Massive venlafaxine overdose resulted in a false positive Abbott AxSYM urine immunoassay for phencyclidine. *J Toxicol Clin Toxicol*. 2003;41:999–1002.

77. Shannon M. Recent ketamine administration can produce a urine toxic screen which is falsely positive for phencyclidine. *Pediatr Emerg Care*. 1998;14:180.

78. Ly BT, Thornton SL, Buono C, Stone JA, et al. False-positive urine phencyclidine immunoassay screen result caused by interference by tramadol and its metabolites. *Ann Emerg Med*. 2012;59:545–547.

79. King AM, Pugh JL, Menke NB, Krasowski MD, et al. Non-fatal tramadol overdose may cause false-positive phencyclidine on Emit-II assay. *Am J Emerg Med*. 2013;31: 444. e5-9.

80. Liu CH, Wang HY, Shen SH, Chiu YW. False positive ketamine urine immunoassay screen result induced by quetiapine: a case report. *J Formos Med Assoc.* 2017;116:720–722.
81. Lotz J, Hafner G, Röhrich J, Zörntlein S, et al. False-positive LSD drug screening induced by a mucolytic medication. *Clin Chem.* 1998;44:1580–1581.
82. Röhrich J, Zörntlein S, Lotz J, Becker J, et al. False-positive LSD testing in urine samples from intensive care patients. *J Anal Toxicol.* 1998;22:393–395.
83. Citterio-Quentin A, Seidel E, Ramuz L, Parant F, et al. LSD screening in urine performed by CEDIA® LSD assay: positive interference with sertraline. *J Anal Toxicol.* 2012;36:289–390.
84. Pearson SD, Ash KO, Urry FM. Mechanism of false negative urine cannabinoid immunoassay screens by Visine eye drops. *Clin Chem.* 1989;35:636–638.
85. Wu A, Bristol B, Sexton K, Cassella-McLane G, Holtman V, Hill DW. Adulteration of urine by urine Luck. *Clin Chem.* 1999;45:1051–1057.
86. ElSohly MA, Feng S, Kopycki WJ, Murphy TP, et al. A procedure to overcome interferences caused by adulterant "Klear" in the GC-MS analysis of 11-nor-Δ9-THC-9-COOH. *J Anal Toxicol.* 1997;20:240–242.
87. Dasgupta A, Wahed A, Wells A. Rapid spot tests for detecting the presence of adulterants in urine specimens submitted for drug testing. *Am J Clin Pathol.* 2002;117:325–329.
88. Cody JT, Valtier S, Kuhlman J. Analysis of morphine and codeine in samples adulterated with Stealth. *J Anal Toxicol.* 2001;25:572–575.
89. Valtier S, Cody JT. A procedure for the detection of Stealth adulterant in urine samples. *Clin Lab Sci.* 2002;15:111–115.
90. Wu A, Schmalz J, Bennett W. Identification of Urin-Aid adulterated urine specimens by fluorometric analysis [Letter]. *Clin Chem.* 1994;40:845–846.
91. Fu S. Adulterants in urine drug testing. *Adv Clin Chem.* 2016;76:123–163.
92. Dunne JR, Tracy JK, Scalea TM, Napolitano L. Lactate and base deficit in trauma: does alcohol or drug use impair predictive accuracy? *J Trauma.* 2005;58:959–966.
93. Thompson WC, Malhotra D, Schammel DP, Blackwell W, et al. False-positive ethanol in clinical and postmortem sera by enzymatic assay: elimination of interference by measuring alcohol in protein-free ultrafiltrate. *Clin Chem.* 1994;40:1594–1595.
94. Rainey P. Relation between serum and whole blood ethanol concentrations. *Clin Chem.* 1999;39:2288–2292.
95. Bertholf RL, Johannsen LM, Bazooband A, Mansouri V. False-positive acetaminophen results in a hyperbilirubinemic patient. *Clin Chem.* 2003;49:695–698.
96. Polson J, Wians Jr FH, Orsulak P, Fuller D. False positive acetaminophen concentrations in patients with liver injury. *Clin Chim Acta.* 2008;391:24–30.
97. Dasgupta A, Zaidi S, Johnson M, Chow L, et al. Use of fluorescence polarization immunoassay for salicylate to avoid positive/negative interference by bilirubin in the Trinder salicylate assay. *Ann Clin Biochem.* 2003;40:684–688.
98. Broughton A, Marenah C, Lawson N. Bilirubin interference with a salicylate assay performed on an Olympus analyzer. *Ann Clin Biochem.* 2000;37:408–410.
99. Tilbrook LK, Slater J, Agarwal A, Cyriac J. An unusual cause of interference in a salicylate assay caused by mitochondrial acetoacetyl-CoA thiolase deficiency. *Ann Clin Biochem.* 2008;45:524–526.

FURTHER READING

1. Courtwright DT. The controlled substances act: how a "big tent" reform became punitive drug law. *Drug Alcohol Depend.* 2004;75:9–15.

Index

Note: Page numbers followed by "f" indicate figures, "t" indicate tables.

Printed in the United States
By Bookmasters